INSIDE INTERNATIONAL TRADE POLICY FORMULATION

A History of the 1982
US–EC Steel Arrangements

by
Michael K. Levine

PRAEGER

PRAEGER SPECIAL STUDIES • PRAEGER SCIENTIFIC

New York • Philadelphia • Eastbourne, UK
Toronto • Hong Kong • Tokyo • Sydney

382.45669
LL6L

Library of Congress Cataloging in Publication Data

Levine, Michael K.
  Inside international trade policy formulation.

  Includes index.
  1. Steel industry and trade--Government policy--
United States.   2. Steel industry and trade--
Government policy--European Economic Community
countries.   3. United States--Foreign economic
relations--European Economic Community countries.
4. European Economic Community countries--Foreign
economic relations--United States.   I. Title.
HD9516.L48   1985   382'.45669142'0973   85-3724
ISBN 0-03-003368-3 (alk. paper)

Published in 1985 by Praeger Publishers
CBS Educational and Professional Publishing, a Division of CBS Inc.
521 Fifth Avenue, New York, NY 10175 USA

Printed in the United States of America on acid-free paper

INTERNATIONAL OFFICES

Orders from outside the United States should be sent to the appropriate address listed below. Orders from areas not
listed below should be placed through CBS International Publishing, 383 Madison Ave., New York, NY 10175 USA

Australia, New Zealand
Holt Saunders, Pty. Ltd., 9 Waltham St., Artarmon, N.S.W. 2064, Sydney, Australia

Canada
Holt, Rinehart & Winston of Canada, 55 Horner Ave., Toronto, Ontario, Canada M8Z 4X6

Europe, the Middle East, & Africa
Holt Saunders, Ltd., 1 St. Anne's Road, Eastbourne, East Sussex, England BN21 3UN

Japan
Holt Saunders, Ltd., Ichibancho Central Building, 22-1 Ichibancho, 3rd Floor, Chiyodaku, Tokyo, Japan

Hong Kong, Southeast Asia
Holt Saunders Asia, Ltd., 10 Fl, Intercontinental Plaza, 94 Granville Road, Tsim Sha Tsui East, Kowloon,
Hong Kong

**Manuscript submissions should be sent to the Editorial Director, Praeger Publishers, 521 Fifth Avenue,
New York, NY 10175 USA**

Acknowledgments

Special thanks to my wife, Holly Lindsay. Most authors acknowledge the patience of a spouse during the writing of a book, but I must thank Holly for her patience during the events of this book. Few mothers-to-be have to make four phone calls to pull their husbands out of post-midnight negotiations when the baby is on the way. I also thank Lynn Holec and Gary Horlick, Commerce's director of the steel import program and deputy assistant secretary for import administration respectively, who gave me the opportunity to stretch my talents to their fullest.

Contents

Acknowledgments v

List of Tables and Acronyms ix

Introduction xi

1 Pre-TPM: US Steel Imports Grow, and Two VRAs 1
  Steel Imports into the US Prior to 1968 1
  The 1968 Steel VRAs 2
  US Steel Imports, 1969-1972 3
  The 1972-74 VRAs 5
  US Steel Imports, 1972–1974 5
  The Consumers Union Law Suit 6

2 The TPM Era 9
  The 1977 Crisis 9
  The Solomon Plan 12
  The First TPM in Operation: May 1978-March 1980 13
  Antidumping Cases and the Birth of TPM II:
   March-October 1980 14
  The Troubled Life of TPM II:
   October 1980-January 1982 17

3 The Beginning of the 1982 Dispute: January-May 1982 26
  New Elements in the 1982 Steel Trade Dispute 26
  The 1982 AD and CVD Petitions 27
  Initial Positions: Carry Through the Investigations 29
  The Investigations Begin: The Initiation
   Decisions 33
  Preliminary Injury Determinations 34
  The Investigations Proceed: Questionnaires
   and Verifications 35

4 Negotiations Prior to the Preliminary CVD
 Determinations: May-June 1982 37
  Baldrige Opens Negotiations: May 1982 37
  Pipe and Tube 41
  Specialty Steel 42

Round One of Settlement Negotiations:
May 19-June 9, 1982   44

5   Preliminary CVD Determinations and Subsequent
        Negotiations: June and July 1982   55
            The Preliminary CVD Determinations: June 10, 1982   55
            Commerce and the EC Move, But US Industry
                Remains Aloof: Early July 1982   58
            Suspension Agreements: Late July 1982   60

6   Commerce Takes a Stand: The August 5 Agreement   65

7   More Commerce Determinations   70
            The Preliminary AD Determinations: August 9, 1982   70
            The Final CVD Determinations: August 24, 1982   71

8   The US Industry Responds: August 25, 1982   74

9   The End of the Game: September-October 21, 1982   79
            Product Coverage and Levels, and Consultations   79
            Enforcement   89
            Article 8: The Shortage Clause   90
            Transfer   92
            Transition Period   92
            Immunization   94
            Pipe and Tube   99
            The EC Approves the Arrangement   109

10   Summary and Conclusion   111

Appendixes   A   August 5 Arrangement   115
             B   EC and US Letters of October 21   125
             C   October 21 Arrangement   129
             D   Pipe and Tube Arrangement   139
             E   CEO Letter   145
             F   Memo on "Price Increase"   147
             G   Minute on Transition Period   148
             H   Request Letters: Section 626   149
             I   Antitrust Letters   151

Index   159

About the Author   165

List of Tables and Acronyms

1-1	US Imports of Iron and Steel, 1957-1968	2
1-2	Comparison of ECSC and Japanese Steel Export Quotas to US Imports, 1971	4
1-3	Comparison of ECSC and Japanese Steel Export Quotas to US Imports, 1972-1974	6
2-1	The EC Steel Market, 1970-1978	10
2-2	Profits of US Steel Industry, 1970-1977	10
4-1	US Market Share of Imports of Basic Steel Mill Products, Total and EC, 1965-1981	40
4-2	EC Import Penetration, Seven Product Categories of June 3 Proposal, 1977–1981	50
6-1	EC Market Share of Stainless Steel Sheet, Strip and Plate, 1977-1981	68
7-1	Comparison of Preliminary to Final CVD Rates, Selected Companies	72
8-1	Comparison of August 25 US Industry Proposal to August 5 Arrangement	76
9-1	US Industry September 29 Proposed Restraint Levels Compared to August 5 Arrangement Levels, Selected Products, Percent of US Market	82
9-2	Estimated Levels of EC Exports of Arrangement Products to the United States, US Apparent Consumption of Those Products, and EC Market Share, August-October Various Years (net tons and percent)	95

LIST OF ACRONYMS

AD:	antidumping
AISI:	American Iron and Steel Institute
AP:	arrangement products
BF:	Belgian franc
CEO:	chief executive officer
CSP:	certain steel products
CU:	Consumers Union
CVD:	countervailing duty
DISC:	Domestic International Sales Corporation
DOC:	Department of Commerce
EC:	European Communities
ECSC:	European Coal and Steel Community
FF:	French franc
GATT:	General Agreement on Tariffs and Trade
IL:	Italian lira
ITC:	International Trade Commission
MFA:	MultiFibre Agreement
OECD:	Organization for Economic Cooperation and Development
R&D:	research and development
TPM:	trigger price mechanism
USTR:	United States Trade Representative
VRA:	Voluntary Restraint Agreements

Introduction: An Overview of the 1982 US–EC Steel Arrangements

Under two steel trade arrangements reached between the United States government and the European Communities (EC) in October 1982, the EC will restrain exports of certain steel products for three years. The completion of the two arrangements capped 15 years of uncertainty in US–EC steel trade, a period that saw the widest variety of trade regimes.

This study recounts the story of how the 1982 arrangements were reached. The events can be read on many levels: industrial policy, trade policy, negotiation, the world steel industry, and the operation of the US government. I present the events as I saw them and interpreted them—from the inside of the US Department of Commerce (DOC)—in the hope of clarifying as much of the history as I can. The full story will require similar contributions from other people in the US government, the EC and its member states, and the US and foreign steel industries.

I will first review the events leading up to the 1982 steel dispute: two voluntary restraint agreements (VRA), followed by the trigger price mechanism (TPM), antidumping (AD) investigations, and a renovated TPM. Next, the final collapse of the TPM is recounted, which leads to the heart of this book: the negotiation of the 1982 US–EC steel trade arrangements.

1　Pre-TPM:
US Steel Imports Grow,
and Two VRAs

STEEL IMPORTS INTO THE UNITED STATES
PRIOR TO 1968*

Prior to 1959 the United States was a net exporter of steel. The war-devastated economies of Europe and Japan required steel to rebuild, but their steel industries were run-down and bombed out. As those industries rebuilt, European and Japanese import requirements tapered off and steel was available for export once again.

In 1959 the US steel industry suffered through a 116-day strike. Imports tripled that year to over 4 million tons to make up the shortfall in domestic production. Each third year when the steel labor contract was under negotiation, imports surged because of hedge buying by steel consumers worried about a strike. These periodic surges were particularly annoying to labor. The availability of foreign steel undermined the extent to which a strike could paralyze the US economy, and made even the threat of a strike costly to the workers (who were laid off even if there was no strike while imports ordered in anticipation of a strike arrived and swollen inventories were worked off).

While labor problems gave foreign producers a foot in the door to the US market by allowing them the opportunity to demonstrate sufficient technical qualifications and reliability as suppliers, other factors sustained

---

*Much of this background material on US steel imports prior to 1972 is taken from *The 1970s: Critical Years for Steel*, William T. Hogan, Lexington Books, Lexington, Mass., 1972.

the growth of US steel imports: the attractiveness of the US market, which is the largest, most open steel market in the world (especially during the period of overvaluation of the dollar, until 1971), and US consumers' demand for low-priced steel. Table 1-1 demonstrates the import trends.

TABLE 1-1.   US Imports of Iron and Steel, 1957-1968 (net tons)

Year	Tonnage Imported
1957	1,154,831
1958	1,707,130
1959*	4,396,354
1960	3,358,752
1961	3,163,233
1962*	4,100,000
1963	5,446,326
1964	6,439,635
1965*	10,383,021
1966	10,753,022
1967	11,454,502
1968*	17,959,886

*Labor contract negotiation year.
Source: American Iron and Steel Institute.

The sharp increase in imports in 1965 (from 6.4 million tons in 1964 to 10.4 million tons in 1965) primed the US domestic political scene for a push to limit imports. The even greater increase in 1968 resulted in the Voluntary Restraint Agreements of that year.

THE 1968 STEEL VRAs

Following an intense lobbying effort by steel companies and labor, Senator Hartke of Indiana introduced a steel quota bill in late 1967. The justification for quotas advanced by the American Iron and Steel Institute (AISI) at the time was the importance of the steel industry to national security, unfairly low wages abroad, and government subsidization of foreign producers. As congressional support grew, foreign steelmakers became anxious. In the summer of 1968 the Japanese and the German steel producers approached the chairmen of the House Ways and Means and Senate Finance Committees. The Japanese and Germans said they would place a voluntary restraint on their exports to the United States and would try to get other countries to do likewise if the Congress shelved the quota bill. An agreement was reached in principle, and the details were worked out by Anthony Solomon, the assistant secretary of state for economic affairs, over a period of months.

The 1968 steel VRAs were entered into by foreign producer associations, not governments. Under the agreements, European Coal and Steel Community (ECSC)* and Japanese producers undertook in 1969 to limit exports to the United States of steel products to about the 1966-1968 average tonnage levels, with 5 % growth per year allowed for the three-year duration of the agreement. They also agreed to maintain product and geographical mix. The restraint was explicitly contingent on the following three assumptions: the restraint did not infringe any US or international laws (this clause was inserted in large part to protect against allegations of violations of US antitrust law); total US imports would not exceed 14 million tons in 1969 and 5 % more in each of 1970 and 1971 (that is, third countries that did not sign the agreement would comply with it); and the US government would not impose any additional steel trade barriers.

US STEEL IMPORTS, 1969–1972

In 1969 US steel imports fell to almost exactly the 14-million-ton objective of the VRA. Imports from Japan fell over one million tons, but still exceeded the export quota by almost half a million tons, an excess explained by the Japanese as late 1968 shipments arriving in early 1969. Imports from the EC fell 1.8 million tons to 5.2 million tons, well below the restraint level.† It is impossible to be certain how much of the drop in imports in 1969 was due to restraint; US steel imports probably would have dropped even without the VRA because of high world steel demand outside the United States, a 5% drop in US demand, and the first year of a three-year labor contract.

---

*The ECSC was formed in the late 1950s and has since been merged with the European Economic Community and the European Atomic Energy Community into the European Communities. This volume uses ECSC and EC interchangeably.

†Because of the time lag between export, import, and the reporting of official statistics, comparison of *export* quotas to US *imports* for calendar years cannot be precise.

In 1970 total US steel imports fell once more; the drop in tonnage from the ECSC was due to a boom in demand in Europe in the early part of the year. In the last four months of 1970, European business slowed considerably and strenuous efforts were made to boost steel exports to the United States. These efforts did not register in increased US imports until early 1971.

By October 1971 US imports already exceeded the 15.4-million-ton VRA goal, and by the end of the year a record 18.3 million tons of steel were imported. It seems clear from import data that both the ECSC and Japanese producers broke their export quotas (see Table 1-2). The ECSC claimed that President Nixon's imposition of a 10% import surcharge had violated the VRA assumption that no additional import barriers would be imposed, and that therefore there was no ECSC violation of the VRA; the US government had destroyed the arrangement itself. While technically correct, the ECSC excuse ignored the three- to four-month lead time between order and import. The surcharge was imposed on August 15; by October 31, only 11 weeks later, the quota had been violated by steel that must have been shipped before August 15 (and sold long before that date). In addition, the weeks immediately after August 15 were marked by great confusion in import markets and few orders were being placed.

TABLE 1-2.   Comparison of ECSC and Japanese Steel Export Quotas to US Imports, 1971 (millions of net tons)

	Export Quota	Actual Imports	Percent Over Quota
ECSC	6.34	7.17	13
Japan	6.34	6.91	9

The 18.3 million tons seems to represent a violation of the VRA in response to strong US demand (1971 was the third highest US apparent consumption in history, up 5.6% from 1970), an imbalance in EC supply and demand (1971 ECSC apparent consumption fell 5% while capacity grew 6%), and customer stockpiling during another labor contract negotiating year.

THE 1972–1974 VRAs

The negotiations to extend the VRA centered on the rate of growth, the inclusion of fabricated structural steel, and specific limitations on specialty steel. The US industry wanted to reduce the 5% rate of growth, to include fabricated structural steel products (imports of which producers of fabricated steel believed had increased because of the VRA restrictions on structural steels), and to protect specialty steel from "quality creep" (foreign producers subject to carbon steel VRAs shifting their exports to more valuable specialty steel). The ECSC faced some internal difficulties, as the Italians wanted a larger share of exports.

The ECSC quota (now including the United Kingdom) was set at 8,013,794 net tons for 1972 (5% above the 1971 VRA level), while the Japanese quota was set at 6,498,059 net tons (2.5% above the 1971 VRA level). The ECSC growth rate for 1973 was set at 1%, for 1974 at 2.5%, while the Japanese growth rate was 2.5% throughout. Thus, while the ECSC had a larger initial quota than Japan, it had a smaller growth rate. Specialty steels were limited to specified tonnages for both the EC and Japan. Cold-finished bars, a non-ECSC product, were specifically limited in 1972 to 2.5% more than 1970 levels, with 2.5% increments thereafter. The Japanese accord provided a stronger geographical mix statement than the 1968 accords; no more than one-third of Japan's exports to the United States would go to Pacific Coast ports. The ECSC geographical commitment remained vague.

The Japanese and the ECSC agreed to limit fabricated structural steels in 1972 to 2.5% more than the 1970 levels, with 2.5% increments thereafter. The ECSC and Japan agreed that no more than 60% of any year's total would be shipped in any calendar semester.

The negotiations for the second VRA foreshadowed many elements that were to be present in negotiations for the steel trade arrangements a decade later: US specialty steel producers' fear of diversion into specialty steel under carbon-only restraint; US cold-finished bar producers seeking specific limitations; concern about next-stage processed products such as fabricated structural steel products; and internal ECSC wrangling over export shares.

US STEEL IMPORTS, 1972–1974

In late 1972 worldwide steel demand began picking up, reaching a peak in late 1973 and 1974. The steel problem in the United States became not too much low-priced imports, but rather a steel shortage and complaints of price gouging. Because of the increase in demand outside the United States, and price controls in the United States, there was no incentive for foreign producers to fill their quotas in 1973–1974 (indeed, US producers increased their exports from 1972 to 1973 by 41% and from 1973 to 1974 by an additional 44%). As Table 1-3 demonstrates, the second VRA did not play an important role in regulating steel trade.

TABLE 1-3.   **Comparison of ECSC and Japanese Steel Export Quotas to US Imports, 1972–1974 (thousands of net tons)**

	1972	1973	1974
ECSC			
Export quota	8,013	8,093	8,295
Actual imports	7,779	6,508	6,424
Japan			
Export quota	6,498	6,660	6,826
Actual imports	6,440	5,637	6,159

The 1972–1974 VRAs were not tested in the marketplace as the 1968–1971 arrangements were; the test of the later VRA came in court.

THE CONSUMERS UNION LAW SUIT

Consumers Union (CU), by an amended complaint filed in October 1972, challenged the legality of the steel VRA. CU's original complaint charged violation of the law on two counts: that the antitrust laws of the United States were violated, and that the president had no authority to negotiate such an arrangement.

Consumers Union soon dropped its allegation that the VRA violated US antitrust law. Two explanations for this action have been offered. The first is that respondents agreed not to challenge CU's standing if the antitrust count was dropped; the second is that CU decided to avoid the multiyear, voluminous proceedings involved in an antitrust case.

Ruling only on the allegation that the secretary of state had reached beyond his legal grasp in negotiating the VRAs, Judge Gesell of the US District Court held that

> the Executive is not preempted and may enter into agreements or diplomatic arrangements with private foreign steel concerns so long as these undertakings do not violate legislation regulating foreign commerce, such as the Sherman Act, and . . . there is no requirement that all such undertakings be first processed under the Trade Expansion Act of 1962.*

Gesell thus ruled that the laws that Congress had passed providing procedures under which industries affected by imports could be given protection did not "blanket the field"; there remained scope for presidential action outside of those statutory procedures.

Gesell noted that the VRAs were initially challenged by plaintiff as a violation of the Sherman Act, but that this contention was dropped. Gesell nevertheless stated in dictum that

> the Executive has no authority under the Constitution or acts of Congress to exempt the Voluntary Restraint Arrangements on Steel from the antitrust laws and . . . such arrangements are not exempt, [and the president] clearly has no authority to give binding assurances that a particular course of conduct, even if encouraged by his representatives, does not violate the Sherman Act or other related congressional enactments any more than he can grant immunity under such laws.†

Gesell's ruling was appealed; the Federal Circuit Court of Appeals for the District of Columbia agreed with Gesell that the negotiation VRA was not an unauthorized executive action, but it expressly disavowed Gesell's dictum on possible antitrust violations (with one dissent).†† The Court of Appeals decision did not put antitrust fears to rest. At the urging of foreign steelmakers and governments, the Department of State requested that Congress pass a retroactive antitrust exemption for the steel VRAs. Congress

---

Consumers Union of US, Inc. v. *Rogers,* 352 F. Supp. 21319, 1323 (DDC. 1973).

†*Id.* at 1323.

††*Consumers Union of US, Inc.* v. *Kissinger,* 506 F.2d 136, 141 (D.C. Cir. 1974).

accommodated this request in the Trade Act of 1974.* The results of this litigation were closely studied by the participants in the 1982 negotiations, and the final shape of the October 1982 arrangements owes much to discussion of the issues raised by Consumers Union—and a concern that Consumers Union or some other party might raise similar issues in court again.

---

*Public Law No. 93-618, §607; 19 U.S.C. §2485.

2   The TPM Era

THE 1977 CRISIS

Steel imports were not much of a problem in the shortage years of 1973 and 1974 nor in the recession that followed. Nevertheless, in 1976 AISI brought a petition under Section 301 of the Trade Act of 1974. It complained that Japan's agreement to limit steel exports to the EC under the Davignon plan had led to increased sales in the United States (Section 301 gives the president broad authority to take action against trading practices he finds to be unfair). After waiting for a year, the US government decided the evidence did not support such a claim.

In 1977 US steel imports jumped 5 million tons to a record 19.3 million. Some 3.5 million tons of the 1977 increase was from the European Community, where stagnant or falling demand and growing capacity were culminating in the steel crisis that continues to the present. Table 2-1 shows the main outlines of the 1977 European steel crisis.

Between 1974 (the peak production year in Europe) and 1977, EC steel production fell 22% (20 million metric tons) while capacity rose 9% (10 million tons)—following a 21% capacity increase (20 million tons) between 1970 and 1974. Little wonder that US steelmakers suspected and accused EC firms of dumping when EC steel exports to the United States doubled in 1977.

Despite an increase in US apparent consumption in 1977 of about 7% from 1976, some US steelmakers came under severe financial pressure: profits disappeared, a calamity after years of subpar performance, as shown in Table 2-2. Only in 1974 had steel's return on equity exceeded the manufacturing average.

9

**TABLE 2-1.   European Community Steel Market, 1970–1978*:
Production, Production Potential, Apparent
Consumption, and Capacity-Utilization Rate
(millions of metric tons and percent)**

	1970	1971	1972	1973	1974	1975	1976	1977	1978
Production	74.2	70.4	76.5	85.0	90.6	68.3	73.5	70.8	74.2
Prod. Pot.	95.9	102.1	107.5	112.1	115.9	121.6	125.2	126.2	129.5
App. Cons.	64.3	60.3	63.1	69.4	67.8	56.6	64.6	61.1	60.3
Cap. Util.(%)	77.4	68.9	71.1	75.9	78.2	56.2	58.7	56.1	57.3
Total Exports to US†	4.1	6.5	5.9	5.0	5.3	3.2	2.4	5.5	6.2

*ECSC products. Includes EC Six only.
†All products.

Source: ECSC Annual Survey of Investment in Community Coal Mining and Iron and Steel Industries, and ECSC Yearbook.

**TABLE 2-2.   Profits of the US Steel Industry and Comparison
to All Manufacturing Average, 1970–1977**

	1970	1971	1972	1973	1974	1975	1976	1977
Profits after taxes (million $)	532	563	775	1,272	2,475	1,595	1,337	22
Profits as % of revenues	2.8	2.8	3.4	4.4	6.5	4.7	3.7	0.1
Profits as % of equity	4.1	4.3	5.8	9.3	17.1	9.8	7.8	0.1
All manufacturing average, % profit on equity	10.1	10.8	12.1	14.9	15.2	12.6	15.0	14.9

Source: American Iron and Steel Institute, Citibank N.A.

Beginning with Gilmore Steel's antidumping complaint against Japanese steel plate filed in March 1977, the industry undertook a campaign for protection from unfair imports. The focus of attack on imports had shifted since 1968: rather than pursue a purely political solution through Congress, the industry (encouraged by the Carter administration) attempted to use the administrative remedies to unfair trade. These remedies had been substantially strengthened in the intervening decade (in part due to the efforts of US steelmakers), and therefore offered more promise of success. In addition, the national concern with inflation (reflected in the economic policies and public statements of both the Ford and Carter administrations) made it unlikely that an appeal against unfair low wages abroad would make much headway, and national security was not a popular issue either. As usual, the industry sought to use the avenue most likely to gain it protection, which in 1977 seemed to be an appeal to equity and a call that laws on the books be enforced.

Few economists have anything but contempt for the antidumping laws currently applied by the United States and other countries. The law's roots are in early antitrust theory. At the beginning of this century, price discrimination by a foreign producer between its home market and the US market with intent to injure a US industry or monopolize trade was made illegal, with criminal and civil penalties.* Because of the difficulty in proving intent, a new administrative approach was soon tried: reducing the remedy and eliminating the need to prove intent. Dumping is now defined as the simultaneous presence of two factors: sales at less than fair value, and injury to a domestic industry from those less than fair value imports. Fair value is normally the price of the subject merchandise in the home market, or in certain circumstances its fully allocated cost of production. If an investigation (normally begun only by petition) finds injury and less than fair value sales, the duty is increased to offset the dumping.

While no discernible economic goal is served by this simplistic concept (Should prices be uniform throughout the world even when exchange rates are in flux, national tastes differ, and the marketplace is ever changing?), the unfairness of dumping in steel was clearer than in other products because the EC ran a steel cartel of sorts, protecting its market, keeping its prices high. United States producers saw lower-priced sales to their country as an unfair attack to which they could not respond without government help. Antidumping petitions seemed the best way to get that help. As for

---

*Act of Sept. 8, 1916, Ch. 463, §801, 39 Stat. 798 (codified in 15 U.S.C. §72 (1970)).

Japan, some US producers did not yet comprehend the extent to which greater efficiency had allowed the Japanese to undersell them with better quality as well, and thought something *must* be unfair.

By late 1977 most of the major US producers had filed, singly or in combination, antidumping complaints. These complaints covered a large proportion of Japanese and European steel imports. The pressure on the government to act mounted as Alan Wood Steel Corporation (with 3,000 employees) went bankrupt and closed its only plant, Youngstown Steel and Bethlehem Steel closed parts of plants in three populous states, and a Congressional Steel Caucus of more than 250 members was formed. The caucus reflected and voiced the industry's doubts that the Department of Treasury (the agency responsible for administering the antidumping statute before 1980) would enforce the law properly, and began considering the legislated imposition of unilateral quotas.

THE SOLOMON PLAN

The Carter administration feared processing the AD complaints to a conclusion for a number of reasons (although it also strongly opposed the unilateral imposition of quotas). The Council on Wage and Price Stability had prepared a report showing that while Japanese firms did enjoy a substantial cost advantage over US producers, European firms did not; therefore, it seemed likely that the European dumping cases would succeed.

To complete the investigations and impose additional duties could, it was thought, interfere with the Tokyo Round multilateral trade negotiations then under way, exacerbate inflation, and perhaps influence the French government's electoral battle against the left. The United States searched for a middle ground and found it in the Solomon Plan, a creative structure masterminded by the negotiator of the 1968 VRAs: Anthony Solomon, who in 1977 was under secretary of treasury for monetary affairs.

The Solomon Plan was announced by President Carter on December 6, 1977 as his policy to "help revitalize the health of the domestic steel industry," "encourage its modernization," and "assist workers, firms, and communities disadvantaged by the depression in the domestic steel industry." The plan addressed domestic issues as well as the allegations of unfair trade, but the steel trigger price mechanism was undoubtedly its centerpiece. The TPM was an immediate success in inducing US steel companies to withdraw their AD complaints soon after the 1977 TPM became operational.

The domestic measures included in the Solomon Plan were implemented simultaneously with the TPM: Treasury modestly liberalized the depreciation schedules for steel equipment; the Environmental Protection Agency reviewed its regulations; the Economic Development Administration issued $365 million in loans and loan guarantees to steel companies and granted loans to depressed steel communities; and the Steel Tripartite Advisory Committee was established. These nontrade measures were more important as symbols of support for the domestic steel industry than as means to dramatically improve its condition.

THE FIRST TPM IN OPERATION: MAY 1978–MARCH 1980

The TPM, which became fully operational on May 1, 1978, soon moderated the fierce price competition among importers. Published trigger prices (for each of thousands of basic steel products) were based on Japanese steel production costs, generally thought to be on average the world's low-cost supplier to the US market. Therefore, the trigger price (adjusted for transport costs) was presumed to be the lowest price at which foreign steel could enter the United States and be at or above fair value, as defined by the AD statute. Steel entering at a price below the applicable trigger price was likely to be priced below its fair value, and could serve as the basis for government-initiated AD investigations. In fact, since the trigger prices in general represented the most efficient producers' fair value, much of the steel entering at or even above the applicable trigger price, particularly from Europe, was presumably priced below its fair value. The TPM implicitly assumed that if steel entered at or above Japanese fair value, any injury or threat of injury from imported steel sold at dumped prices would be substantially eliminated.

Japanese steel production costs were converted to US dollars for trigger prices by using an average of the daily dollar/yen exchange rate for the 60 days preceding the quarterly TPM calculation of production costs. The designers of the original TPM did not anticipate the large fluctuation in the dollar/yen exchange rate through 1978 and 1979. During 1978 the yen appreciated rapidly, causing the TPM 60-day yen/dollar average to decline from 240 yen/dollar for the original trigger price calculation to 187 yen/dollar for the calculation of first-quarter 1979 trigger prices.

This exchange rate movement alone increased trigger prices by 18.9%, which, added to increases in Japanese production costs, resulted in a total 21.8% increase in trigger prices during TPM's first year. Trigger prices would have increased 3% more in the first quarter of 1979 but for Treasury's use of the discretion built into the first TPM, which allowed it to establish trigger prices 5% higher or lower than the actual production cost estimated.

During 1979 the yen/dollar exchange rate reversed its 1978 trend, causing the TPM 60-day yen/dollar average to fall to 227 yen/dollar for the first-quarter 1980 trigger price calculation (a 21% drop in one year). This depreciation of the yen more than offset increases in Japanese production costs. As a result, trigger prices, even with Treasury using its 5% discretion band to keep them slightly higher than Japanese production costs, actually decreased slightly between the first and fourth quarters of 1979. Since domestic steel production costs increased throughout 1979, the decline in trigger prices meant that domestic producers found it increasingly difficult to compete with steel imported at trigger prices. The domestic industry became frustrated with the failure of trigger prices to rise with US costs, which effectively decreased import protection at a time of falling domestic steel demand. In addition, the industry was disillusioned with the failure of the Carter administration to support either improved tax incentives for capital formation or substantial revisions of environmental requirements. US Steel Corporation abandoned support of the TPM in March 1980 and filed antidumping complaints against steel producers in seven European countries.

During the lifetime of the first TPM, only two steel antidumping investigations were self-initiated by the Treasury Department, and those were against minor participants in the US market (Poland and Taiwan). Importers selling below trigger prices were not a major reason for the first TPM's demise; quite simply, the US industry refused to put up with trigger prices based on Japanese costs once those costs fell too far below its prices. And changes in US trade law since TPM's inception provided the industry the weapon it needed to seek "improvements" to the TPM.

ANTIDUMPING CASES AND THE BIRTH OF TPM II: MARCH–OCTOBER 1980

Following the successful conclusion of the Tokyo Round of Multilateral Trade Negotiations in 1979, Congress enacted the Trade Agreements Act of 1979, which among other actions, implemented the results of the agreements. The conclusion of the Tokyo Round removed one of the most compelling impediments to US government prosecution of antidumping complaints against the EC, while changes in US law (the result of intensive lobbying efforts by domestic steel producers, among others) greatly increased the prospect of success of antidumping and countervailing duty (CVD) cases against foreign producers. The decisions of the executive branch and the International Trade Commission (ITC) were made subject to judicial review, time limits on investigations were substantially tightened, and the administration of the AD and CVD laws was transferred from the Treasury Department, which was not trusted by Congress or domestic producers to properly enforce the laws, to the Commerce Department.

The steel imports included in US Steel's March 1980 petitions accounted for about 26% of the steel imported into the United States in 1979 and $1.4 billion in trade. US Steel alleged dumping margins ranging from 12% to 45%; if margins of this or similar levels had been found and the ITC had found injury or threat of injury, the majority of this steel would have been shut out of the market. Upon US Steel's filing of these cases, the TPM was suspended, since the Treasury Department had announced that it was designed to substitute for, not supplement, individual company dumping petitions.

The EC Commission, recognizing the likely outcome of the AD proceedings, requested reinstatement of the TPM, with at least the implied threat of pressures on exports to Europe of certain US goods. The Carter administration was sympathetic to the EC's pleading, and Commerce and the US trade representative (USTR) began intense discussions with the domestic steel industry to determine precisely the reasons for the industry's loss of faith in the original TPM—with an eye toward changing the system to allow its reinstatement. The three big problems with the TPM were identified as: the variability of price protection due to yen fluctuations; the lack of domestic industry understanding of how trigger prices were calculated and how the system was enforced; and the potential for injury from dumping, even though sales were made at or above trigger prices.

Considerable governmental pressure was placed on US Steel (including members of its board) to withdraw its antidumping petitions, even though some or all of the cases would probably have resulted in positive findings. Along with changes to the TPM, these pressures succeeded, and in October 1980 the cases were withdrawn and the second TPM put into effect. The main features of the revised TPM were: a dollar/yen exchange rate conversion base equal to the average rate of the past 36 months rather than 60 days (which boosted trigger prices immediately by 12%); increased transparency in TPM methodology and procedures; and provisions for responding to surges in imports of specific steel products even if priced above trigger prices (but perhaps below fair value) if US capacity utilization was not high and imports were.

Other characteristics of the reinstated TPM included an understanding with the domestic industry and with the European Commission that the new TPM had been reinstated for a period of either three or five years, depending on the results of the secretary of Commerce's assessment of the domestic steel industry's modernization efforts. If the secretary deemed those efforts inadequate, he could terminate the TPM after a three-year run.

Finally, recognizing the growing potential for disputes over EC steel subsidies to disrupt the TPM, the second TPM was explicitly expanded to cover unfair subsidies. Producers were protected from harm due to subsidization of foreign competitors by the countervailing duty law, which works similarly to the AD law: when subsidization and injury are proved by an investigation, duties to offset the subsidy are imposed. The CVD law was quite undeveloped because it had been little used in the past, and rarely had it been applied to the domestic (as opposed to export-linked) subsidies involved in the European steel industry. But its potential to disrupt US–EC steel trade had grown to rival the dumping law because, for various reasons, the British, French, Belgian, and Italian governments had poured money into their steel industries to help them through the crisis of the late 1970s. The amounts of money involved were in some cases staggering: a German Steel Association study published in February 1981 conservatively estimated the subsidies given in these four countries between 1975 and 1981 at over $10 billion.

The expansion of TPM to cover subsidization as well as dumping was not carefully thought out. Rather, it was done out of necessity, for if it had not been done, US firms would have been free to file CVD petitions while enjoying TPM protection (and thereby defeating TPM's primary purpose of maintaining peace with the EC).

The only mention of TPM's expanded purview in the notice of its reinstatement (45 F.R. 66833, October 8, 1980) was the listing of the following as one of the four important premises on which TPM II was based:

> The TPM is designed to promote the elimination of *injurious dumping and subsidization* of imported steel and thereby to moderate the adverse effects on the domestic industry that can result from such unfair import competition. [emphasis added]

It was not specified just how TPM II would deal with subsidies; hindsight reveals that this murkiness was costly, as EC producers either did not understand or chose to ignore the implications of TPM's expansion. The ultimate collapse of TPM II was largely attributable to its inability to address the subsidization issue in a timely and coherent fashion.

THE TROUBLED LIFE OF TPM II: OCTOBER 1980–JANUARY 1982

The period following TPM's reinstatement was the first time that trigger prices were in place during a substantial world slump in demand. The effects of this slump on prices led to tensions that ultimately destroyed the TPM for a second time.

During the first two years of TPM's existence, trigger prices had been slightly lower than US prices, which in turn were, it seems fairly certain, lower than European producers' home market prices or costs. By selling at trigger prices, foreign producers were able to maintain the slight price differential to compensate US consumers for longer delivery times, the non-cancelability of orders, and so forth, and did not have to sacrifice much (if any) volume. The major effects of the TPM in its first incarnation on foreign producers were: to enable them to remain in the US market despite probable technical violation of the AD and CVD laws; and by reducing price competition, to increase prices earned from the sale of steel in the United States over pretrigger levels.

The US demand for steel, and consequently steel prices, was quite low when trigger prices were reinstated in October 1980, and world demand— particularly EC demand—was just beginning to weaken. Given the state of US prices and demand, foreign producers were concerned about the reinstatement of trigger prices (particularly at the higher level). But optimism about future demand and the pressure of the pending preliminary dumping determination (due October 17, 1980) caused the Europeans to press for TPM's reinstatement.

While overall US demand for steel improved in the first half of 1981, demand and prices for some major products, notably sheet products (which go into autos and appliances for new houses, sales of which were severely depressed), remained *below* trigger price levels. At the same time that overall US demand (temporarily) strengthened, EC steel demand collapsed, and competition among EC producers, always lurking behind various official and semiofficial restraint schemes, broke out again. The price competition in the EC was reportedly fierce, as the heavily subsidized producers cut prices to boost production and avoid politically difficult layoffs, seemingly without regard to accumulating losses.

As a result of the relative strength of US demand and the sharp depreciation of the European currencies (from October 1980 to July 1981 the German mark depreciated 24% against the dollar; the Dutch guilder, 28%; and the French franc, 29%), United States steel prices appeared extremely attractive to European producers, who could sell in their depressed home markets only at a loss.

The EC steel producers, tempted by the relatively attractive US market (they said they could make more money selling in the United States than in Europe, even net shipping costs), had a problem: the TPM, which in the past had been so beneficial to them. To be competitive in the US market, they said, they could not price their steel above US producers' prices, which implied they had to sell below trigger prices. This, for a while at least, they were reluctant to do, for fear of destroying the recently negotiated compromise on TPM II. However, as the pressure to export more to the United States grew, EC steelmakers approached the US government seeking help. Trigger prices were above both US and EC prices and EC cost of production, they said, and some provision should be made to allow EC sales in the United States without forcing them to undersell trigger prices.

The US government had faced the problem of trigger prices being above non-Japanese producers' fair value as early as 1978. Then, Canadian producers had claimed that because of the efficiency of their plants and their transport cost advantage over Japan, they could sell steel in the United States below trigger prices and not be dumping. In response to Canadian requests, the Treasury Department had established a "preclearance procedure" under which foreign producers could submit to an informal investigation establishing that their fair value was below trigger prices, thereby removing the risk of provoking AD investigations by below-trigger sales. The procedure had been simple and the investigations fairly cursory; ultimately, all Canadian producers had been exempted from TPM I. Recognizing Canadian concerns, the preclearance program was reactivated along with TPM II. Initially, preclearances were taken to be as they were under TPM I: a stamp of approval from the US government that a foreign producer could sell below trigger prices without selling below "fair value" in terms of the *dumping* law. No explicit recognition of TPM's expansion to the realm of subsidies was made (in practical terms, preclearance investigations did not at first examine possible subsidization).

Several EC firms saw the preclearance program as a possible answer to their TPM difficulties and sought preclearance soon after TPM II's inception. These firms included Sacilor and Usinor, which had received substantial amounts of money from the French government, Hoogovens of the Netherlands, and a few large German firms. Commerce, in the midst of the Carter-Reagan transition, had not yet established a firm plan on how to deal with EC firms' requests, and sent mixed signals. On the one hand, Commerce offered preclearances to anxious EC firms as a solution to their pricing concerns; on the other, Commerce (and the EC) saw grave problems connected with preclearance (that is, too many preclearances would provoke US steelmakers into filing a new round of petitions). However, EC firms could not wait for a resolution between the EC and US governments and pushed ahead their applications. Some firms reportedly held back on exports to the United States pending a decision on their applications; others decided to risk the consequences of continued, even boosted exports. Rumors circulated that at least one EC producer was using its preclearance application to tell US buyers concerned about its low prices that the prices would be approved by the forthcoming preclearance.

Commerce moved slowly on the European preclearance requests, mindful of the potential consequences of granting too many preclearances. In February 1981 it decided to require preclearance applicants thought to be subsidized to complete detailed subsidy questionnaires, which answers would be taken into account in an unspecified manner in the preclearance decision. This decision, affecting primarily the French, was meant as much as anything to scare applicants out of pursuing preclearances, and to delay the preclearance decisions. However, even the apparently heavily subsidized French continued to press their applications, on the assumption that any subsidy was small compared to the true cost advantage the strong dollar provided them. In addition, for several months the TPM bureaucracy was being rebuilt and was badly understaffed (preclearance investigations were essentially as complex as dumping investigations). As Commerce struggled over policy on preclearances, on rebuilding the bureaucracy, and in familiarizing the new Reagan administration with TPM's intricacies, developments in steel trade began to make Commerce's delaying tactics increasingly inadequate.

United States steel imports from the EC were low in the first quarter of 1981, down 16% (149,000 tons) from the first-quarter 1980 low level (total US steel imports were down 10%). The decline in sheet imports, on which the EC's large integrated firms relied most heavily, was huge: down 43% from the first quarter of 1980. The drop in sheet imports was due in part to weak US demand. However, reluctance to sell below trigger prices without a preclearance cost some firms dearly—notably Hoogovens, the Dutch sheet steelmaker, from whom US imports fell to zero in the first quarter of

1981 from 48,000 tons in the first quarter of 1980. Total steel imports from the EC would have fallen much more but for buoyant pipe and tube sales (usually made by other producers than the sheetmakers).

As it became clear that difficulties were developing, the EC Commission became very concerned with what its representatives called the "imbalance" between trigger prices and US market prices. It saw a number of possible consequences, all of them bad: EC firms would refuse to sell below trigger prices and their exports would not recover from dismal first-quarter levels; some EC firms would receive preclearance while others would not, resulting in politically difficult-to-handle differential treatment among EC producers; DOC would preclear too many firms and provoke the US industry to repudiate the TPM and file AD and CVD complaints; or EC firms would, one way or another, ignore or avoid the TPM and boost exports with unknown reaction from the US government and steel industry. Consequently, beginning in February 1981 the commission urged the new administration to lower trigger prices at least 10% and suggested a number of ways to do so (for example, use a Canadian rather than a Japanese cost base). With lower trigger prices, EC firms could participate in the market without "violating" the TPM.

A few days after commission officials left Washington in May 1981, having argued strenuously for a cut in the trigger prices, April import figures became available. They showed a big increase in both total US imports (from 1.14 million tons in March to 1.76 million tons in April) and in US imports from the EC (from 227,000 to 558,000 tons). It both relieved and worried DOC officials that sheet imports from the EC increased 60,000 tons—relieved because the EC might quit pressuring for alterations in the TPM, worried because the increase could mean EC firms had decided to sell below trigger prices with abandon. By May it was clear that some EC producers were ignoring trigger prices. For example, most US imports of hot-rolled sheet from France between January and July 1981 were more than 10% below trigger prices.

DOC officials understood the difficulty faced by EC firms and considered the commission's suggestions. However, since TPM II was only months old, it was believed that any major changes—and certainly a 10% cut—in the carefully negotiated structure would cause the very result TPM was designed to prevent: the filing of AD and CVD complaints. DOC struggled with the fundamental question of how TPM, a system designed to address antidumping concerns, could be made relevant to both EC and US producers when the focus of the debate had shifted to subsidization (most informed observers believed that the appreciation of the dollar had wiped out any dumping margins that might have previously existed).

In June 1981 Commerce, under new secretary Malcolm Baldrige, developed a plan to address TPM's problems. It decided to beef up enforcement of the TPM so that Europeans would not feel they could ignore it and the US industry would not feel compelled to file petitions of its own. The alternatives were few; the only other plans discussed were to let the TPM die a well-deserved death and take whatever cases came, or to replace the TPM with some sort of antisurge mechanism.

Baldrige's plan was simple. DOC would self-initiate investigations, including CVD cases, when it thought both subsidization or dumping *and* injury were present. In short, Commerce would monitor imports, prepare cases to be brought, and pick off the most flagrant violations of the trade laws. In a sense, trigger prices became almost irrelevant, except to help identify the most blatant violators. Commerce's staff would essentially assume the role of monitoring imports and foreign subsidization and preparing petitions normally done by the steel firms themselves, with the added role of trying to use the threat of cases to jawbone foreign steelmakers into more temperate behavior. Baldrige hoped that by announcing this policy and showing determination to carry it through, foreign producers would be deterred from engaging in unfair trade and US producers would refrain from filing their shotgun cases.

Baldrige revealed part of his plan at a June 26 press conference featuring the self-initiation of three minor antidumping cases (nails from Japan, Korea, and Yugoslavia) and some actions to improve the (suspect) integrity of trigger price monitoring information (audits of importers related to foreign producers, and pursuit of fraudulent invoicing). Baldrige also explicitly stated that no preclearance would be granted to any firm that had received significant subsidies. This position meant that the threat of French preclearances, for example, was past, but by the same token French and other subsidized producers could not look to preclearances for relief from the TPM imbalance.

To implement Baldrige's plan, DOC undertook an examination, with input from domestic steel producers, of steel subsidies in certain countries in preparation for the initiation of cases. DOC warned certain firms that they faced potential self-initiated CVD cases if increased imports presented a convincing injury case, even if declining currencies protected them from self-initiated AD cases. At the same time, Baldrige authorized the hiring of 30 more staff for the department's import administration; the additional help would be needed either to beef up enforcement of the TPM or to work on the flood of cases that would result if TPM broke down.

In parallel with preparation to monitor more closely US imports for unfair trade and prepare and prosecute cases, DOC held discussions with domestic steel companies—including face-to-face meetings between Secretary Baldrige and several chief executive officers—to probe the limits of the industry's tolerance for modifications to the TPM that would make trigger prices a valid reference price once again. DOC determined that its maximum ability to lower trigger prices without provoking a slew of CVD petitions (for which industry drafts had already been prepared) was TPM's historic 5% flexibility band.

In discussions beginning in July 1981 between the US government and EC Commission staff and between the EC Commission and its constituent firms, a possible TPM compromise was explored. This compromise would have, in essence, established an effective import price floor for sheet sales in the Great Lakes (the most important product market for the large EC producers) at prices 5% lower than second-quarter trigger prices, with the concurrent step of abolition of preclearances (which would relieve the EC Commission of a sticky dilemma). The level of uncertainty in the market would have been reduced and the intergovernmental wrangling quieted. The proposal contained obvious problems. The EC Commission would have had to police compliance with the lower trigger price levels—that is, enforce the minimum price system from its end (although it is not clear that the commission had the power to do so if the firms were not cooperative). The compromise fell through when it became apparent that a 5% decrease in trigger prices was not enough, in the opinion of the Europeans, to allow them to continue to sell sheet in the United States. The EC producers said at least a 7% cut would be needed, while the US producers, who were sullen about a 5% cut, would have rebelled at any decrease of more than 5% and filed cases. The European firms and the EC apparently decided it was better to endure the uncertainty of a tattered import price floor and unknown US steel industry and government reaction than to forego remunerative below-trigger sales.

Shortly after the discussion over reducing trigger price levels broke down in mid-September, the import statistics for August 1981 (reflecting sales in spring 1981) became available, showing the highest steel imports in US history, much of it from the EC. This all but guaranteed AD or CVD cases, for two reasons: the industry was furious, reasoning that the enormous volume of spring sales showed that EC producers had taken advantage of it while talks had been going on; and the sharp rise in imports meant that affirmative injury findings (necessary for AD or CVD cases) were now possible. DOC, seeking to forestall across-the-board filings by the industry and the destruction of the TPM, moved to selectively self-initiate cases (as had been promised publicly when TPM II was announced). Seven cases were chosen. Baldrige announced these in early November before the Senate Steel Caucus. Significantly, these cases included two countervailing duty cases involving EC countries: hot-rolled sheet from France and plate from Belgium. Other cases involved Romania, Spain, South Africa, Canada, and Brazil.

In addition, Commerce eliminated the preclearance program entirely. It had completed some of its preclearance investigations and saw no way, short of eliminating the program entirely, to avoid granting preclearance to Hoogovens, Korean pipe manufacturers, and others. Given the current agitated state of the US industry, Commerce officials wanted to avoid further provocation—enough to outweigh the desire to accommodate those foreign producers the TPM unfairly penalized.

Many observers, including the EC and EC steel firms, apparently had believed that the US government would never initiate CVD investigations on its own. It had, under the Treasury Department, made every effort to avoid enforcing the CVD statute as it applied to domestic subsidies because of the political sensitivity of attacking foreign governments' economic policies and the fear of opening up a Pandora's box by ruling on previously unexamined potential subsidies (such as, government ownership of industry). Commerce initiated the five CVD investigations reluctantly, but felt it had no choice if TPM was to maintain any credibility at all—which was essential if there was to be any chance to avoid a massive onslaught of cases.

Commerce's selective case initiation/deterence approach was not accepted by US steel producers. They felt they were being injured on a broader front and believed they had a good chance of winning a far larger number of cases. In addition, they were angry with EC producers and so sought broader penalties, and they mistrusted the EC producers and so sought certainty. By December steel company threats to file a large number of AD and CVD complaints (either to get higher duties on most foreign producers or to force a quota system) appeared very credible, as imports remained high, domestic demand plummeted, and US capacity utilization fell below 60%.

On December 4, 1981, President Reagan met with the heads of several US steel firms. He asked the firms to delay their complaints until Secretary of Commerce Baldrige had a chance to talk to the Europeans one more time. The CEOs agreed. Baldrige journeyed to Europe and explained to EC officials that their producers' disregard of trigger prices and massive sales of steel in the United States had undermined the TPM. It was up to them to devise a solution that would assure US steelmakers that TPM could remain an effective method of preventing injurious dumping or subsidization. At the same time, the administration was reluctant to reach any overt voluntary restraint agreement, in view of its overall commitment to free trade, except where a violation of the unfair trade laws was found (that is, "free but fair trade").

In late December EC officials presented their plan to Baldrige. It consisted of a system of forecasts and reports of aggregate shipments, which the EC would make available to its producers, who could then adapt their sales programs to avoid disruption of the US market. It was left unclear (probably deliberately) whether this implied export quotas, to be enforced by the EC Commission if necessary. In any case, there was no explicit or implied commitment that EC steel exports to the United States would be reduced; in fact, the EC stipulated that if it was to implement its proposed system, DOC would "administer TPM in such a way so as not to inhibit traditional trade flows."

Baldrige did not endorse the EC system, but agreed to convey the proposal to the US industry, along with the first forecast of a 10% reduction in exports of EC products in 1982 (from 1981). While the US industry considered the plan, news arrived of the latest round of EC subsidy authorizations under the State Aids Code (an agreement adopted by the EC, at German urging, which aimed at eliminating subsidies to EC steel producers). The commission authorized the Belgian government to convert 5.2 billion BF (about $140 million) of Cockerill-Sambre debt to equity and to extend an additional 4.1 billion BF ($110 million) loan to the ailing firm. The commission also granted 850 million BF ($23 million) to Cockerill-Sambre for investment projects. France was authorized to put an additional 4.43 billion FF ($816 million) into its steel industry, and Italy was allowed to add 350 billion IL ($308 million) in equity to its industry. The timing could not have been worse.

On January 7, 1982, William Delancey, the head of AISI, responded to the EC's proposal. He and his colleagues had a number of questions about the system, including, inter alia, whether the projections were actually commitments, how the system would work beyond 1982, how the forecasts would be drawn up, and how the EC would sanction violators. The essence of the problem was that the industry wanted absolute certainty—either denial of entry to imports below trigger prices (which was conceded to be impossible without an unlikely change in US tariff law), or enforceable quotas. The industry would not settle for less than that certainty, because it felt it had been cheated twice by the TPM (which had not proven to be the guaranteed minimum import price they had wanted), and because it felt it could win most if not all of the more than 100 cases it planned to file. While some EC producers and officials were probably amenable to quotas ("traditional patterns of trade"), it is unlikely that most felt threatened enough in December 1981 to agree to enforcement sanctions. Even had the EC been willing to reach a purely political agreement, the US administration was not. Both because of its free-trade commitment and because senior officials did not feel the industry would accept reasonable quotas as long as it thought it could win all of its cases, the US government did not want to negotiate steel quotas in December 1981.

While the industry hoped that US–EC talks would continue, Delancey said it could not wait until these questions were resolved to file its trade complaints; if the talks did not reach an acceptable agreement, the industry would be damaged by the delay. Four days later, seven steelmakers filed a total of 132 AD and CVD complaints. As the US producers had been warned, TPM was suspended as to all products and all countries (Commerce timidly refrained from terminating it—TPM II remains in a state of suspended animation to this date), and the investigatory process begun. The EC's December 1981 proposal was not pursued further.

In retrospect, it is now clear that TPM II failed largely because it was fighting the last war (dumping) instead of the new war (subsidies). When the government realized the difficulty, it was too late. By late summer 1981, when EC steelmakers let loose of all restraint and sent a flood of steel into a sharply declining market, TPM II lost all credibility and had to be replaced.

3   The Beginning of the 1982 Dispute: January-May 1982

NEW ELEMENTS IN THE 1982 STEEL TRADE DISPUTE

Although the filing of cases and suspension of TPM in January 1982 superficially resembled the filing/suspension in March 1980, the events of the intervening two years had caused many changes. The US steel companies had become convinced that any price-monitoring system was unworkable because of evasion by increased sales through related parties (even though, in fact, related-party transactions remained at about the same level from 1979 through 1981) and now sought either increased tariffs or firm quotas. Organizationally, too, the industry position changed: in 1980 US Steel had been the only petitioner, and thus the only party with a chance to negotiate for withdrawal of its petitions. With TPM's failure, other steel companies wanted a seat at the negotiating table for its replacement. Thus, seven major firms filed petitions.

The seven firms divided into three groups: US Steel, represented by in-house counsel; Bethlehem Steel, the second largest firm, represented by an antidumping/countervailing duty law specialist; and "the Five" (Republic, National, Inland, Jones & Laughlin, and Cyclops), who jointly hired a large New York firm. In 1982 any negotiated settlement had to please all the major steel firms, not just the biggest one. And as nonintegrated firms saw their TPM protection withdrawn, they filed cases as well (for example, wire rod producers and rail producers).

The US industry was not the only fragmented side. The EC was also divided by differing interests because of the new prominence of the subsidy issue. The Germans and Dutch had little fear of countervailing duties *if* the US government administered its law in a reasonable manner (which could not be taken for granted, as most of the issues being considered were novel and complex), while it was virtually certain that high subsidies would be found in France, Belgium, Italy, and the UK. The Germans in particular had been vehemently protesting steel subsidies in other EC countries (as German firms had no import protection within the Common Market) and had pressed the EC into adopting a weak commitment to phase out state aid by 1985.

Finally, the 1982 negotiations—and it was perceived as likely from the time the January 1982 petitions were filed that there would be negotiations—would be driven by an extremely complex and unprecedented legal process. The 1977 TPM introduction and the 1980 reintroduction were to some extent driven by the pending dumping cases, but in neither instance did the administrative proceeding go beyond the initiation of investigations. By the time Baldrige opened discussions to try to reach a negotiated settlement through quotas in May, the preliminary CVD determinations were only a month off and could not be avoided. From there on, each statutorily mandated deadline for the cases became a negotiating target and a turning point.

The statutory timetable mandated a Commerce determination of whether the petitions were a sufficient basis to initiate investigations by February 1; ITC preliminary injury determinations by February 25; preliminary CVD determinations (with attendant contingent liability for duties) by June 10; and preliminary AD determinations by August 9. Final CVD determinations would be due August 24, which meant July 24 was the last day for a suspension of the investigations if an agreement under the CVD statute was reached. Final AD determinations would not be required until December, with ITC final injury determinations ending the investigation process 45 days after publication of Commerce's final determination.

THE 1982 AD AND CVD PETITIONS

The three groups coordinated their petition filing only to the extent of all filing on the same day; the petitions varied dramatically in most other respects. The sloppiness and unreasonable subsidy allegations (supported by patently ridiculous notions of subsidy quantification) of some of the petitions enhanced foreign suspicions that the US industry was harassing imports rather than seriously pursuing unfair trade. In total, 132 separate cases were filed, of which 96 were against EC producers and nine each against South Africa, Brazil, Spain, and Romania. (A case is defined as an AD or CVD investigation of a product from a country, for example, an AD investigation of plate from the UK.) Ninety-four of the cases were CVD and 38 were AD; 66 consisted of AD and CVD cases against the same product/country.

The original round of petitions (more came in the following months)* was directed primarily against EC producers. Spain and South Africa had been included because they had not signed the General Agreement on Tariffs and Trade's (GATT) Subsidy Code and therefore did not receive an injury test, while the petitions against Brazil and Romania responded to particularly egregious low-priced import surges of specific products. The most important non-EC cases were already under investigation because of the cases self-initiated by DOC under the TPM, which were terminated and reinitiated with the steel companies rather than the Commerce Department as petitioner.

The petitions alleged both dumping and subsidization. The dumping allegations were generally based on a comparison of selling prices of imported steel to cost of production constructed from published annual financial reports. United States officials were skeptical that the allegations of dumping would be substantiated, given the very strong dollar (from 1980 to the date of the preliminary dumping determinations, the simple average appreciation of the dollar against the relevant currencies was 47%), but were required to initiate investigations by the terms of the statute. Once the investigations began, the data provided by the petitioners were discarded—except as a threat if the foreign companies refused to cooperate by providing detailed data. Because dumping investigations had been quite common and most of the firms had been through an investigation less than two years before, there was little controversy at the start.

The same cannot be said for the CVD cases. Because there was little previous case experience with domestic subsidies, the petitioners attempted to frame the issues broadly, labeling every conceivable government program a subsidy and offering techniques to quantify the subsidies. The less-prepared petitioners relied heavily in some instances on research done by the Commerce Department for the TPM-initiated cases, but some petitions were the result of extensive (and expensive) research and analysis. The petitions were obviously self-interested documents, and some EC officials and steelmakers rightly ridiculed the more glaring errors and outrageous claims. However, data on foreign subsidization are not easily available and the economic, financial, and legal analyses were quite difficult. On balance, the petitions clearly offered a basis for investigation.

---

*Between January 11 and November 30, 1983, at least 50 additional steel cases were filed with the Commerce Department.

The alleged subsidies included grants, low-interest loans, forgiveness of debt, low-cost raw material supply, various labor subsidies, R&D subsidies, government equity participation, export subsidies, and more. It was incontrovertible that there was much foreign government intervention in steel industries. There were two questions. First, how would Commerce determine which foreign government policies constituted subsidies and how would those subsidies be quantified for the purpose of applying offsetting duties? Second, would ITC find that the allegedly subsidized imports—which accounted for 20% of total US steel imports in 1981 (45% of imports of the specific products involved) and 3% of total US apparent steel consumption (8% of consumption of the specific products)—were causing material injury to US steelmakers?

INITIAL POSITIONS: CARRY THROUGH THE INVESTIGATIONS

Upon the filing of petitions on January 11, all parties stated they wanted to see the investigations through to conclusion rather than construct another agreement along the lines of the TPM or a voluntary restraint agreement. The US steel industry felt it had good cases. Even if Commerce or ITC found against it at times, judicial review of the negative decisions would maintain uncertainty in the steel trade for years to come. After the filling, the industry devoted its resources to the cases (at some cost to detailed preparations for a possible negotiated settlement), although David Roderick, the chairman of US Steel, did say he was encouraged when Baldrige announced that US–EC discussions would continue after the petitions were filed.

The Reagan administration stated that it intended to push the investigations through to conclusion, vigorously enforcing US trade laws which, it claimed, were consistent with international trade law. Baldrige had found the TPM a headache to administer and was not sorry to be rid of it. The Reagan administration generally preferred to allow the market the predominant role in allocating resources; the preferred steel policy was to provide an atmosphere of competition, unpolluted by subsidies and other unfair trade practices, and allow the US steel industry to find its own equilibrium. In addition, the administration had a no-lose position domestically if it could claim it had firmly and effectively enforced existing trade laws. Baldrige had determined early on that he would not be stampeded by voluminous industry complaints, and so he had provided sufficient money and as many staff positions as could be spared to ensure the department could investigate as many complaints as the industry could file. Commerce staff was ready to go when the petitions were filed.

Nevertheless, because of expectations that strict enforcement of unfair trade laws would prove too costly in terms of US–EC relations, Baldrige kept open the option of reaching a negotiated settlement of the dispute. He stated his position succinctly in a March 23 interview with *American Metal Market:*

> Let's clear the smoke first through the pursuance of the complaints. I'm convinced that both sides think they are right. Rightly or wrongly, the Europeans don't think that they are subsidizing or dumping their steel. The Americans say they are. In the past this has been negotiated, and although I don't preclude the chance of a negotiated settlement I think the best way to go is to settle it—take it all the way and find out if the laws are being violated.

The EC's initial response to the filing of the petitions was a mixture of anger and determination to prove themselves innocent. Most of the EC steel producers already had been through AD cases in 1977–1978 and 1980, and they prepared to defend themselves before the Commerce Department and the ITC. The EC, through Vice-President and Industry Commissioner Davignon, went a step further and began a public defense aimed at both the European and US populations. Because Davignon's two primary arguments were repeated continually over the course of the investigations, it is worthwhile to consider them here.

First, Davignon maintained that the difficulties faced by the US industry were the result of the collapse of domestic demand and that imports from the EC were not to blame. According to this interpretation, the EC was being scapegoated. In statements on January 9 and 13, 1982, Davignon noted that 1981 US steel demand was 20% less than in 1979, and that the EC share of the US market fell from 6.7% in 1979 to only 4.7% in 1981. He further asserted that sales of European steel in the US market showed a greater decrease in 1981 from 1979 (-16%) than the drop in both steel production and consumption in the United States (-12%). According to Davignon, EC producers had suffered more from the drop in demand than US producers themselves had, rather than exacerbating the decline in demand as US producers claimed.

Later in the year (August 13), the *Washington Post* offered the steel negotiators this advice: "Don't get entangled in the numbers." In the steel dispute, there was little *but* numbers, and Davignon's first public argument tried to use numbers to make his case that there was no injury. Commerce's response was twofold: first, to remind the EC and state publicly that the injury determination is made following a careful, detailed investigation by the ITC, an autonomous body, on a product-by-product basis (comparisons of total steel figures are of little analytical value). Second, Commerce staff tried to replicate Davignon's numbers, failed, and then phoned EC staff for explanation. The explanation, after discussions between DOC and EC experts, illustrated the possibility of each side convincing itself it was right. It turned out that the supposed 1979 figure of 6.7% EC market share was actually from 1978, a peak year for the EC, and was limited to the ECSC steel definition, which excludes pipe and tubes, where EC export increases had been largest (but which were not at the time the subject of cases). The EC's actual 1979 market share (by their definition of steel products) was 5%, while the comparable 1981 share was 4.9% for the first ten months. In addition, the EC had used ten-month 1981 data, rather than available 11-month data (full-year 1981 data were not yet available); because of the rising trend of imports from the EC and the falling trend of US demand, the 11-month EC market share for 1981 had risen to 5.2%—above the comparable 1979 figure.

The same import surge in late 1981 also explained the EC's figures of 12% US market decline and 16% EC import decline—they were based on ten-month data. Adding in November 1981 revealed a 14% decline in US consumption from 1979 and only an 11.4% decline in imports from the EC.

This detailed excursion through the numbers demonstrates the ease with which "the numbers" could be manipulated. It also dramatically illustrates what caused the filing of the petitions: a massive surge in imports in the second half of 1981. The surge was undeniable, and the ITC preliminary determinations of injury would be based less on historical yearly comparisons than on the effects of that surge in the marketplace in the last half of 1981.

Davignon's second argument was that the entire world steel industry was in a crisis, that all countries in the Organization for Economic Cooperation and Development (OECD) had recognized this in 1977 and agreed "on the need for a profound restructuring of the sector and on the principle that the sacrifices needed would be shared equitably without threatening the traditional trade flows." Davignon seemed to imply that since the EC was well within its traditional market share, any measures taken by the United States against the EC would be a violation of this agreement.

This assertion that the EC had a *right* to its traditional US market share permeated EC pronouncements and proposals. It was coupled with a rejection of US notions of fair versus unfair trade and a feeling that the EC was being unfairly singled out for persecution. In Davignon's press conference on January 13, he said the United States "ought first to analyze their [sic] imports from Japan and Canada," despite the fact that the industry had uncovered no evidence of unfair trade from those countries.

The reliance on the so-called OECD consensus was, from the US point of view, misplaced. In 1977 an ad hoc OECD group had been convened to consider the world steel crisis and agreed that

> no nation can be expected to absorb for sustained periods large quantities of imports of unjustifiably low prices to the detriment of domestic production and employment, but any measure designed to deal with such imports *should take into account customary patterns of trade.*

It is to the emphasized language that the EC continually referred.

The US retort to the assertion that it would violate the commitments if it enforced its unfair trade laws was threefold. First, it was noted that the TPM, an unusual trade policy measure by any standard, had been implemented to protect the EC from the full application of US trade laws, and that the collapse of the TPM in 1982 was directly caused by European immoderation. Second, the United States pointed to a later OECD declaration, the declaration establishing the OECD Steel Committee in 1978. By that time, US negotiators realized what the EC claimed the 1977 language to mean, and pressed hard for—and got—language changes. The 1978 committee mandate declared that

> When taking action under domestic law and procedures to deal with serious difficulties of its industry, a participant shall take into account *the concerns of its trading partners* that traditional trade flows *established under normal conditions of competition* not be severely disrupted. [emphasis added]

"Normal conditions of competition" was understood by the United States to mean not dumped or subsidized. The fundamental US position was that the EC share of the US market should be determined by a fair marketplace, while the EC wanted its share protected politically. Finally, officials continually noted that US trade laws were consistent with international rules agreed upon as recently as 1979.

A major element of Davignon's public strategy was to separate the US steel industry—which in his mind was intransigent and "whose only goal was to obtain from Europe an auto-limitation of its exports to such a level that it could have led to a drastic decrease in the traditional presence of the European exporters on the US market"—from the US government. Davignon repeatedly stated that he still had respect for and confidence in Secretary Baldrige, pointedly blaming the steel industry for TPM's collapse, rather than Baldrige's refusal to lower trigger prices on his own. The EC, it seemed, would fight the US industry's allegations of unfair trade, proving to Baldrige and the ITC that countervailing duties on Europe's exports could not be justified. FRG Economics Minister Lambsdorff also noted that the dispute was between the EC and the US industry, not the EC and the US government. He stressed that a self-restraint agreement was not a possible solution to the conflict. He was not publicly joined in the latter opinion by representatives of other EC member states.

THE INVESTIGATIONS BEGIN: THE INITIATION DECISIONS

United States law and regulations specify in great detail the procedure for antidumping and countervailing duty investigations. The first phase of both is the Commerce Department's determination, within 20 days after a petition is filed, of whether the petition is a sufficient basis on which to initiate an investigation. Because investigations are expensive for participants and tend to chill trade, initiation decisions are important and can be controversial.

The law requires that an investigation be initiated if an interested party representative of a domestic industry (a producer, labor union, or business association) alleges the elements necessary for the imposition of an AD or CVD duty (normally subsidy or sales at less than fair value *and* injury) and supports those allegations with information reasonably available to it. Following congressional mandate in 1979, the evidence requirement is flexible: for large, wealthy firms like steelmakers, the requirement is stiff, while for small businesses less evidence is required. The decision to initiate implies no judgment on the merits of a case.

The cases represented a major challenge for Commerce's import administration. There were nearly four times more petitions than had ever been submitted before, and they involved novel and sensitive CVD issues. Fortunately, there had been an opportunity to plan ahead, with country teams formed, headed by experienced personnel (totaling 75 analysts), extra copiers and filing cabinets ordered, and a computer program for the dumping cases ready by the date of filing. A superstructure was also created, with an issues group of senior staff and lawyers, and a separate unit of 15 people formed to prepare briefing materials, answer the voluminous correspondence that resulted from the cases, and prepare for possible negotiations (including the input of all available steel statistics into a computer).

Each petition was considered on its own merits. Where supporting data were lacking, petitioners were given the opportunity to supplement. Several petitions were so lacking of evidence that they seemed to invite rejection, and the petitioners were so informed and given the opportunity to withdraw those petitions rather than have them rejected. Most of those cases involved instances of zero or de minimus (less than a fraction of a percentage of the market) levels of recent imports and nothing beyond allegations to indicate a threat of increased future imports. In addition, because there had been no imports nor evidence of offers of Romanian sheet products, no investigation of the dumping allegations could possibly proceed—there was no price into the United States to compare with foreign market value. The petitioners ultimately withdrew 16 petitions, and Commerce rejected seven that petitioners refused to withdraw. On February 1, 1982, 109 investigations were initiated, covering $1.8 billion of steel imports in the first 11 months of 1981, about 20% of US steel imports.

The petitioners whose cases had been dismissed protested in court that Commerce did not have the authority to reject a petition merely because there had been no imports and there was no evidence to indicate a threat of future imports.* Commerce's position was that "the law is not concerned with trifles." The EC, on the other hand, argued that 47 of the 63 EC product/country combinations should have been dismissed on the basis of insignificant imports (37 of these were initiated, five withdrawn, and five dismissed). This request for dismissal included hot-rolled sheet from the Netherlands (1981 imports of 255,000 tons, about 13% of total imports of that product) and cold-rolled sheet from France (imports of 174,000 tons, 12% of total).

PRELIMINARY INJURY DETERMINATIONS

By the forty-fifth day after a petition is filed, the US ITC is required to determine whether "there is a reasonable indication that an industry in the

*A judge at the base-level court upheld the petitioners' claim in this regard. The court cases were withdrawn as part of the ultimate settlement before a government appeal of that decision could be made, leaving the precedential value of the lower court decision unclear.

United States is materially injured or threatened with material injury by reason of imports" allegedly dumped or subsidized. In early 1982 the ITC was composed of five commissioners, appointed by the president and confirmed by Congress. Eighteen investigators, financial analysts, and steel industry experts on the ITC staff set to work on a crash basis for that 45-day period, sorting through the volumes of data and argumentation presented by all parties to enable the commissioners to make their decisions.

Commission decisions are made by majority vote. While each commissioner is required to explain the reasons for a decision, the commission as a whole is not. Thus, it is very difficult to explain in a short space the ITC decisions on the 92 cases before it (under US law, no proof of injury was required in the cases of imports from Spain and South Africa). Each commissioner considered the data on imports and import trends, profitability, lost sales, pricing, uses, and so forth that had been submitted and gathered, and then made an evaluation. One commissioner found injury in all cases, primarily because he saw the entire steel industry being possibly harmed by any unfair trade; another made judgments for entire product groups without regard to the country of exportation; the others judged each case individually.

The result of the ITC voting was the elimination of 54 of the 92 cases considered. However, the 54 cases terminated comprised only 10% of the 1981 imports covered by the original petitions. The ITC eliminated all the unimportant cases, making administration of the cases much more manageable. More important, the elimination of so many country/product combinations reduced the appeal to the US industry of carrying the cases to conclusion (because the exporter could in some cases switch to the products that had been eliminated) rather than getting a more comprehensive negotiated settlement, and showed the EC that the US government was administering the cases fairly.

THE INVESTIGATIONS PROCEED: QUESTIONNAIRES AND VERIFICATIONS

The Commerce Department was determined that the investigations should be seen to be handled thoroughly and fairly. The goal was to be able to issue sustainable determinations if the cases went to a conclusion and, more important, to show both sides that the government was willing and able to complete the investigation if need be. (There was a belief that the government had been forced to give away too much in the 1980 negotiations because of the perceived inability of the Commerce Department at that time—it had just received responsibility for the AD and CVD laws—to complete the investigations in a manner that would stand up to judicial scrutiny.)

The cases were treated as normal AD and CVD cases: questionnaires were prepared and issued to the foreign companies and governments; arguments from domestics and foreigners on the data received were considered; investigators were sent abroad to verify the data submitted by examining company and government records; and the policy issues of determination and quantification of subsidies were researched and considered. Because premature release of information indicating likely outcomes could result in marketplace actions (such as buying/selling stocks or securities, building/ liquidating inventories). Commerce tried to keep its deliberations and subsequent negotiations cloaked in secrecy. It was successful to a remarkable degree. As the investigations proceeded, it became clear to all concerned that Commerce was not being overwhelmed. The investigations would be completed if no settlement was reached.

4 Negotiations Prior to The Preliminary CVD Determinations: May–June 1982

BALDRIGE OPENS NEGOTIATIONS: MAY 1982

Although Commerce and the EC had maintained contact at several levels (Baldrige and Under Secretary for International Trade Lionel Olmer with Davignon and Davignon's chef de cabinet, Hugo Paemen; Deputy Assistant Secretary for Import Administration Gary Horlick with his counterpart Gerard dePayre) and Commerce staff had begun background preparation for possible settlement negotiations even before the petitions were filed, there was no direct US–EC negotiation in even the vaguest terms until May. The first public hint of a willingness to negotiate a settlement was dropped extemporaneously by Secretary Baldrige in a hearing before the House Commerce Committee in March 1982, when he stated in passing that he would be willing to accept an agreement if it were acceptable to both sides.

The matter was formally briefed to the Cabinet Counsel on Commerce and Trade in April 1982. A decision was taken to seek a negotiated settlement to the cases, with Baldrige in the lead but in cooperation with USTR Brock. In the first week of May, the government began discussions toward a settlement of the steel trade dispute that would be acceptable to the European Community. All relevant agencies were on board, and it was agreed that Baldrige would conduct the talks for the United States.

A negotiated settlement had become important to both the United States and the EC. At a time when harmony was needed, US–EC relations were bad and seemingly getting worse. The Soviet Union was in Afghanistan, Poland was in turmoil, and the proposed stationing of additional US missiles in Europe and President Reagan's vocal anti-Communism were testing the alliance.

On top of these simmering basic issues facing the alliance, three specific disputes had arisen, seemingly simultaneously. In addition to steel, the United States and EC were at odds over agricultural export subsidies and the Soviet gas pipeline, which the Reagan administration strenuously opposed and was trying to convince Europeans to oppose as well. Steel seemed to be the area where compromise was most likely, where US–EC relations could begin to mend (and where domestic charges that Reagan's militancy was wrecking the alliance could be countered).

Washington was also concerned about undermining the EC. Steel was, along with the Common Agricultural Policy, one of the underpinnings of the communities. Davignon warned in an interview with *Metal Bulletin* (Feb. 2, 1982) that US countervailing duty action, by threatening EC exports, threatened the EC steel policy. Resulting production cuts could be so stiff, he said, that "people might lose their cool." The British had noted that if the US market was cut off, Germany was the logical alternative market, while the Germans threatened to impose border barriers against EC steel if forced to do so. The United States had historically felt that a strong EC was in its national interest, and feared internal EC discord.

Finally, there was the threat of retaliation. While the only publicly stated EC threat concerned the Domestic International Sales Corporation (DISC) (a US export subsidy that had been found to be in violation of the GATT but that the internally divided US government had lacked the political will to change), murmurings of other potential actions abounded—particularly from the French, whose newly elected Socialist government initially saw the CVD cases as an ideological attack on socialist policies. The United States was not overly concerned about EC-wide retaliation because it was confident that so long as the investigations were fairly handled, the Dutch and Germans would block any such attempts.

On May 9, 1982, Baldrige, USTR Brock, and Commerce Under Secretary Olmer met with Vice-president Davignon and other EC officials. The Europeans were told the United States was willing to try to negotiate quotas as a solution to the steel dispute, but that there was no urgency from the US side to do so. Baldrige said in no uncertain terms that the United States was willing to help the EC get out of the mess they had gotten themselves into by greedily blowing apart the second TPM. Baldrige outlined his ideas of what a settlement would look like. The basic concept would be a quantitative restraint agreement; Baldrige, and nearly everyone else involved in steel trade problems, had learned from the TPM the follies of

price-monitoring schemes. Baldrige wanted the agreement linked to EC reduction of subsidies, and stated that if the AD and CVD cases were taken to
their conclusion, he would allow fairly traded EC steel into the US market
(and, by implication, impose the appropriate duties on unfairly traded
steel). He communicated his positions on the major points of a VRA as follows:

- *Product Coverage.* All steel products should be included, including
  pipe and tube, wire rod, and specialty steel. The United States knew the
  EC could most easily control ECSC products (those subject to ECSC
  pricing, production, and other controls), which were primarily produced by major integrated producers but did not include pipe and tube.
  However, Baldrige felt he could not get US industry agreement on a
  less-than-comprehensive VRA, and he was not eager to try to short-cir-
  cuit the investigations against the steel industry's will. In essence, Baldrige felt there was an opportunity for compromise by trading breadth
  of coverage, by products and countries, for the very sharp specific limitations on particular products/countries that the investigations would
  likely bring about. Davignon informed Baldrige that he felt he could
  negotiate export licensing of all steel products.
- *Enforcement.* Baldrige's initial position was that any enforcement of
  the VRA should be from Europe. No easy legal means for US enforcement was available. Nevertheless, because it was foreseen that both
  sides would prefer US enforcement, Commerce Department staff continued to look for an enforcement mechanism.
- *Duration.* Baldrige wanted the agreement to run through at least
  1985—the date the TPM was scheduled to expire, the date at which EC
  steel subsidies would supposedly end, and far enough after the 1984
  presidential election to allow relatively unpressured consideration of a
  follow-on regime.
- *Level of Permitted Imports.* Baldrige had, in previous discussions with
  US producers, explored their ideas about what an acceptable level of
  imports from the EC would be. His reading was that 11.8% of the market was acceptable (a level last seen in 1966, as shown in Table 4-1),
  and that the US industry expected the EC to account for about one-third
  of that total. Baldrige told Davignon that he thought he could "stretch"
  the 11.8% to 13% (a level last seen in 1973), implying an acceptable
  EC market share of all steel producers of 4.33%. The EC, on first hearing Baldrige's proposal, was taken aback at the 13% figure, noting that
  a total minimum import penetration of about 15% was more appro-

priate. In fact, as Table 4-1 shows, 4.33% was lower than the EC market share of all years since 1965 but two (1976 and 1980). When the EC had time to study the data more thoroughly, 4.33% would come to appear unacceptable.

TABLE 4-1.   **US Market Share of Imports of Basic Steel Mill Products, Total and EC,* 1965–1981 (percent of total market)**

Year	EC	Total
1965	4.8	10.1
1966	4.6	10.7
1967	5.9	12.0
1968	7.7	16.4
1969	5.8	13.4
1970	5.6	13.8
1971	8.3	17.8
1972	7.4	16.6
1973	5.3	12.4
1974	5.4	13.5
1975	4.7	13.5
1976	3.2	14.1
1977	6.3	17.8
1978	6.5	18.1
1979	4.8	15.2
1980	4.1	16.3
1981	6.2	19.1

*Includes all EC members as of 1981

Overall, the EC reacted eagerly to Baldrige's presentation. Davignon indicated he would like to negotiate intensely to reach agreement on a VRA before the June 10 CVD preliminary determination. Before plunging into the negotiations between May 9 and June 10, we must consider in more detail two critical issues of product coverage that figured prominently in those negotiations: pipe and tube and specialty steel.

PIPE AND TUBE

Pipe and tube had been a bright spot for US steelmakers in 1982, sustaining them while other product lines suffered through a long recession. Demand had skyrocketed because oil drilling boomed and buyers, fearing shortages, scrambled to buy all the pipe they could get. United States production rose 20% between 1979 and 1982, and prices rose by $200 a ton and more. Imports also benefited, more than doubling in those years to make up for a shortfall in domestic production.

EC pipe and tube producers were primary beneficiaries of high US demand. In 1982 pipe and tube imports from the EC jumped almost six times to 1.8 million tons, taking almost 11% of the market. Prior to 1981, the EC share of the US steel pipe and tube market never reached higher than 1972's 4.6%, and from 1974 to 1981 was under 2% in four separate years.

In early 1982, however, as oil prices stabilized, the market collapsed. A huge inventory overhung the market, steel pipe and tube prices plummeted, and new orders to US producers alost ceased. United States producers were hurt, and they blamed foreign competition—notably the EC producers. The US industry claimed it could have supplied the market with all the pipe and tube it needed; in fact, it said, there had been no shortage. Rather, foreign firms, in large part European, had wrecked the market by shipping in too much product, resulting in disastrous inventory buildup rather than steady high profits. The US firms wanted revenge and protection, and were willing to use the leverage from AD and CVD petitions against other products to gain them. The first shot over the EC's bow came in a CVD petition filed in May 1982 involving only a small portion of EC pipe and tube exports, but alleging subsidies against all the major EC producers.

The EC pipe and tube producers, on the other hand, felt they were innocent of any wrongdoing and were unfairly under attack. The largest EC producer, Mannesmann of West Germany, was a privately owned, profitable, well-respected maker of quality tubes. Its officials maintained that it had simply responded to demand and profit opportunities.

Unlike in the United States, in Europe the steel pipe and tube industry is separate from the integrated steel industry. ECSC regulations do not apply to pipe and steel, and EC pipe and tube makers have jealously guarded their independence. Pipe and tube produced within nationalized steel industries were normally in separate units. The attitude of the EC pipe makers was that they had not dumped in the United States, they were not subsidized, and they were not willing to sacrifice their profits. Their attitude, in short, was "the *carbon* steel industry caused the problem—they should pay, not we."

The EC and the member states were caught in the middle: they could try to force the pipe and tube makers into concessions, at some internal political cost, or they could hold out and hope the US industry would accept an agreement without pipe and tube, a course the US government continually advised against. Baldrige and Olmer were convinced the US industry would never agree to anything that did not include some limitations on EC pipe and tube.

SPECIALTY STEEL

Specialty steel is usually defined as stainless steel plus alloy tool steel. It comprises thousands of separate products, each of which has a base of carbon steel with various chemicals added. Specialty steel is much more expensive than carbon steel (often $2,000 per ton or more, compared to carbon's $350-$600), and is made by different processes in much smaller quantities.

In both the EC and the United States, the carbon and specialty steel industries overlap to some extent. The major integrated producers manufacture some specialty products, and some firms make nothing but specialty steel. In both continents, the producers who make only specialty steel are the more influential in regard to policies affecting specialty steel, but the integrated producers retain some influence. This relationship between specialty and carbon producers gave the EC an influence over specialty steel it did not have in pipe and tube.

The US specialty steel producers had political clout out of proportion to their size in the economy because of their image as efficient high-tech producers, national security concerns, and long, effective lobbying by the industry's members. The Reagan administration wanted to avoid a steel agreement that did not include specialty steel; cries of "Diversion!" could be anticipated from specialty steel producers, who claimed foreign producers would shift exports to specialty steel if carbon steel exports were limited.

Specialty steel producers had first been successful in gaining import protection in 1972, when specialty steel was added to the EC and Japanese VRAs. In 1976 they were given quotas for five years following an investigation under Section 201 of the Trade Act of 1974, the "escape clause" (which allows relief to be granted to firms suffering serious injury from imports—even if those imports are not unfairly traded). When those quotas expired in 1980 (after an extension), the industry wanted to be included in TPM, but did not have any unfair trade cases pending at the time, the withdrawal of which could have been used as a bargaining chip with the Carter administration to obtain that goal. (There are also technical reasons, relating to the number and complexity of specialty steel products, why a separate trigger price for each specialty steel product would have been, at best, very difficult.) The industry's political clout, especially in Pennsylvania, did win an election year promise from Carter that "something" would be done, and (*after* Carter lost the election) a "specialty steel surge mechanism" was devised to "do something" for the industry.

Under the specialty steel surge mechanism, Commerce was to monitor imports and self-initiate AD and CVD cases where appropriate. The standard of appropriateness was never defined, beyond the statement by the relevant government official to the industry that "We'll know it [the appropriate case] when we see it." The surge mechanism was nothing more than a device for jawboning foreign products whose exports were growing, and as such it actually worked for a little more than six months. As specialty steel imports started increasing, following the earlier trend of carbon steel imports, the industry began requesting relief from the government. The differences of interpretation over the surge mechanism then became apparent. From the Commerce Department's point of view, there was much less evidence of subsidization of European specialty steel production than was the case with carbon steel, and evidence of dumping prices was difficult to come by (for the government *and* the industry) because of the highly specialized nature of the product (unlike carbon steel, where prices for standard grades in significant amounts of tonnage are quasipublic). In addition, the specialty steel industry had difficulty assembling the requisite injury information, and Commerce staff was stretched too thin to allow intensive research and preparation of petitions for self-initiation. From the industry's point of view, they had been promised by the Carter administration that they would be protected from imports, and they were not very interested in the details.

When the carbon industry made known its plans to file AD and CVD cases, the specialty steel industry decided it wanted to have cases filed also to avoid being left out of the settlement. The specialty industry, however, was less interested in these cases than was the carbon industry because its European competitors were less vulnerable, and also because it had gotten used to the quotas of 1976-1980 (against all imports) and did not want the selective barriers against some imports implied by separate AD and CVD cases. At the same time, the industry evidently did not feel it could get relief under the escape clause again, in part because of the need for presidential approval in a new administration committed to free trade. Consequently, on December 2, 1981, the specialty steel industry, with great publicity, filed a petition with the USTR under Section 301 of the Trade Act of 1974, alleging many unfair trade practices against major foreign producers—notably subsidies. Section 301 has fewer procedural protections than the AD and CVD laws and is much less expensive, but it also requires presidential approval of any remedy (which might come more readily if legitimatized as protection from unfair trade, rather than simply as protection under the escape clause).

The December petition was deficient in several respects, and the industry was forced to withdraw and refile on February 26. For several reasons, the interagency review groups were unanimous that the revised petition should be rejected: it was poorly documented; the CVD law was the proper place for the complaint; and Section 301 itself was so vague and broad as to be difficult to apply. However, the political reasons against simply rejecting the petition were too strong, and the petition was accepted and an investigation begun. But that only made USTR's situation more difficult—it now had to find a way to satisfy the domestic industry without actually providing any relief (which could not be defended in the GATT). For an administration looking for a way out of the Section 301 investigation it had gotten itself into, the best solution would be to include specialty steel in a broader steel agreement.

ROUND ONE OF SETTLEMENT NEGOTIATIONS: MAY 19–JUNE 9, 1982

On May 19, 1982, Under Secretary Olmer met with Hugo Paemen, Davignon's chef de cabinet, to exchange ideas on how a settlement could be reached. Olmer's ideas were essentially those Baldrige had presented to Davignon ten days earlier, with some additions. A specific link between the "export program" and the EC's restructuring effort was proposed, with each EC country's export level to be dependent on compliance with the EC State Aids Code. Olmer proposed a new, slightly lower level of restraint, 3.95% instead of the 4.33% of ten days earlier. This reduction was proposed because Commerce staff had searched for a logical basis for the 4.33% (which had been arbitrarily selected as one-third of 13, as explained above), and 3.95% was the number closest to 4.33 for which a "logical" basis could be found (the basis was a reduction of EC exports from a certain period by the amount of overcapacity in the EC).

In addition, Olmer for the first time discussed the mechanics of the proposed VRA. Because the US industry was so suspicious of the EC and US governments (remembering the experience with the vagueness of the 1968 VRA), and the EC steel industry also mistrusted the commission and the United States, the agreement would have to be in great detail. This detail would be refined as talks progressed and more minds were applied to the specific language. The general idea Olmer presented May 19 remained the same throughout: the EC would be given a specific market share for each steel product, which share would be applied each year to projected US consumption of that product to arrive at the annual allowed export tonnage. The projection would be reviewed quarterly.

Paemen in turn orally presented EC ideas. He agreed that market shares for each product along with forecasting was the best mechanism, and enforcement through EC export licensing seemed acceptable. He added the first complication, however: an adjustment to EC exports in February of each year to reconcile forecasted and actual US consumption for the prior calendar year. While no specific levels were suggested, Paemen indicated that EC steel exports should be lowered in 1982 and brought back to "normal" levels by 1986—that is, the program would end on December 31, 1985. His initial product coverage included only sheet, plate, and structurals, the biggest categories, although he said other products could probably be added. Pipe and tube and specialty steel would not be able to be covered by June 10.

Paemen presented two EC conditions for agreement: that the United States forbid entry to imports for the EC not accompanied by an export license, and that the share of the market EC producers relinquished would not be assumed by other non-US producers.

The EC's desire for US enforcement stemmed from a fear that its powers alone were insufficient to ensure orderly export control. For example, an EC producer could ship steel to Canada and subsequently reexport it to the United States. If the United States would refuse entry to that steel, such circumvention would be impossible. If such circumvention occurred, the United States would expect the EC to adjust its export limits and punish the violator, two tasks the commission would rather not perform.

The desire of the EC that it not be replaced in the US market by third countries was related to its "traditional market share" theory and its rejection of the US contention that EC steel exports had been in any way unfair.

In talks over the next week, Commerce Department representatives emphasized that no guarantees about third countries could be made except that the United States would apply its trade laws strictly. It was agreed that an independent forecaster would provide projections of US consumption, rather than either government; it was thought that this would leave less room for political argument. No levels could be agreed upon, and indeed little new discussion on the levels occurred. As talks progressed, the EC added two new wrinkles: the question of immunizing EC producers from trade actions for the duration of the VRA and a series of "flexibility" provisions.

The EC focus on immunization was, it seems, in response to EC steel producers' demands. The Germans in particular resented having to periodically defend themselves against what they claimed were ludicrous charges of unfair trade, and wanted at a minimum to receive an assurance that if an agreement were reached the expense and uncertainty of more investigations would be avoided for its duration. Eventually, the EC came to characterize this issue as its desire for "peace in the valley." The US difficulty was that under its law domestic firms have a right to file petitions that cannot be rejected if they meet statutory criteria. The law prevented Commerce from agreeing to what was thought to be a legitimate request.

The flexibility provisions added at this time involved carryover, advance use, and transfer. The first two would allow the EC to use a specified portion of a year's permitted export of a product in either the previous or the following year, while transfer would allow tonnage shifts between product categories. These clauses, EC negotiators argued, were standard in agreements of this sort (in fact, the proposals were taken from standard bilateral textile agreements under the MultiFibre Agreement [MFA]). They were needed to allow some flexibility, in response to the market.

On May 26, just two weeks before the preliminary CVD determinations were due, the EC submitted to the Commerce Department a draft agreement covering most points except for the restraint level. Commerce negotiators were frustrated that despite long discussion on key points, the EC had moved so little from positions that were clearly unacceptable. May 26 had been a target date for the EC Council of Ministers to meet to approve an export program, but the EC proposal remained so far from acceptable that no meeting could occur. The key points of the draft agreement, along with Commerce comments that were immediately conveyed to the EC, were as follows.

First, the EC required that the US government refuse all petitions for relief from unfair trade with respect to covered products for the duration of the agreement. Commerce had no legal authority to do so, and neither desired nor thought it possible to get such authority.

Second, the EC acceded to the demand for wider product coverage by adding coverage of wire rod, bars, and galvanized sheet. With the addition of galvanized sheet, the EC abandoned its position of imposing export restraint on only those products under investigation; the ITC had found "no injury" in all the galvanized sheet cases and terminated those investigations. Pipe and tube and specialty steel remained outside the program.

Third, the EC solidified its proposal that its exports not be supplanted by third countries. The draft stated that

> if the EC share of the combined . . . imports of the seven products covered by this Agreement is found to have been below ⅓ of total US imports of these products, the above-mentioned figure [the export limitation] will be adjusted accordingly.

This was clearly unacceptable as it would require either that total imports of covered products be kept to below three times the EC share, or that the EC share be increased in direct proportion to the increase of other imports.

Fourth, the EC wanted its market shares to increase 5 and 10% in 1984 and 1985 ("in view of progressive implementation of State Aids Code"). Commerce noted that growth would be acceptable if the initial level was low enough. No level was proposed by the EC until June 4.

Fifth, the EC's proposed flexibility provisions were made more specific. Advance use was to be allowed up to 10% of total exports of a product category, carryover to 25%, and transfer to 25%. Commerce officials thought those figures far too high.

Sixth, the EC continued to demand US denial of entry of unlicensed merchandise.

Finally, EC negotiators added another concept: that EC producers be allowed to increase exports to the United States during periods of steel shortages. The concept was acceptable in principle to the Commerce Department. The logic behind this was twofold: to protect US consumers from unnecessarily severe restriction (which would cause problems for US steel producers, who would be painted as the villains of any such restrictions), and to allow the EC to supply some of the imports that otherwise would come from other foreign suppliers.

The EC's short-supply formula was that all restrictions on exports be lifted if the US industry operated at 85% of capacity or more for three successive months. The problem with the language was threefold: it was not product-by-product, meaning that US producers would not accept it (the industry could be above 85% but still have some products in ample supply); the 85% figure was too low, given the high historic operating rates of the US industry; and there was a timing problem—no indication was given of how long the relaxation of export restraint would last.

The next day (May 27) Baldrige met in New York with the CEOs of the major steel companies. He and Olmer had been periodically in touch with the steel chiefs singly by phone and in person, but now they were all gathered together to hear what Baldrige and the EC had to propose. Commerce officials decided against passing out the draft agreement the EC had submitted the day before, on the theory that its more extreme and/or poorly phrased proposals would so anger the US producers as to damage the negotiations. Instead, the CEOs were presented with a description of the proposed agreement: the products the EC would cover, the duration (to 1985), the forecasting mechanism, the *US-proposed* level (3.95% for *all* products, allocated to covered products using a 1979-81 base), the EC's carryover, advance use, and transfer requirements (but with no percentage figures), and an explanation of the government's legal difficulty with enforcement. The CEOs were reportedly annoyed with what was called "the flexibility system"; they saw it as full of gimmicks inserted to circumvent any limitations. They saw no chance that an agreement based on that system could be reached before June 10.

On May 28 Olmer reported on the CEO meeting to Paemen. After hearing what amounted to a confirmation by the CEOs of warnings Commerce officials had previously conveyed on the flexibility provisions, Paemen suggested that talks should end. He expressed his view that the US government was stalling and that further talks would not be useful so long as the government was not willing to force the industry to accept a "fair" solution—as Paeman claimed the EC was doing to its steelmakers. Baldrige then phoned Davignon, who agreed to reassess the EC position and get back to Baldrige as soon as possible.

Only ten days remained until the announcement of the preliminary CVD determination, which the EC so wanted to avoid. Over the weekend of May 30-31, Commerce staff and lawyers for the US steelmakers met to discuss the proposed VRA in detail. It turned out there had been misunderstandings about some aspects of the plan, including the base year by which the EC's total share of the US market would be allocated to specific products. At least one important CEO had understood that 1982 was to be the base year, meaning the EC would be given an unacceptably large slice of the US pipe and tube market (assuming that pipe and tube was ultimately added to the system). The lawyers were forthright with their views: the agreement must last beyond 1985 because of the long-lasting effects of the subsidies, and because the fruits of the US modernization program would not be on-line until 1986; pipe and tube must be covered; the 3.95% market share in total was acceptable, but growth in 1984 and 1985 was not; the flexibility gimmicks were tough to understand, reduced the certainty of relief, and were unacceptable; and US enforcement was necessary. Certain companies mentioned additional products they wanted covered, such as cold-finished bar and alloy bar, and a few said specialty steel should be covered as well.

These talks gave Commerce officials a good idea of the state of mind of the US industry: it was eager to see the cases progress and would accept an alternative only if it met *all* its important demands. Because the VRA proposed to date was, in the view of the industry, so flawed in basic outline, little attention was paid to details. This caused difficulties later.

On June 3, after days of intense internal negotiations, the EC presented its assessment of the situation and another draft agreement. Davignon wrote to Baldrige that as far as the community was concerned, it had been able to establish a framework that would allow negotiation of a final settlement. The reason the EC was comfortable with the framework was that essentially none of the objections of either the US government or the US steel industry had been taken into account.

Product coverage remained limited to seven categories. Pipe and tube could not be included, wrote Davignon, because EC pipe and tube producers considered themselves innocent and would fight charges of unfair trade. Specialty steel posed difficult problems and was not included because the large number of small producers in the EC made export control difficult (that is, allocation of tonnage in quantities large enough to allow profitable sales for individual companies), but some specialty steel could be included if inclusion was an absolute condition.

For the first time, the EC proposed a restraint level: 6.7% for the seven product category total. Davignon asserted that the EC historic share, based on 1977-1979, was more than 7.5%, but that "in a considerable effort I could bring this down to 6.7% on the condition however that all the other elements of the system we suggest be adopted." The EC's 6.7% proposal seemed very high to the US negotiators, as Table 4-2 demonstrates. The EC would have the US industry agree to a level higher than any since 1978, above the levels in 1979 and 1981 that had provoked massive AD and CVD petitions, at a time when the industry had cases pending that it believed would eliminate much EC steel from the US market.

Although the EC's suggested restraint level alone would have made agreement impossible, it should be noted that the "other elements" remained unacceptable to both Commerce and the US industry. Commerce could not legally make a commitment to reject unfair trade petitions or to refuse entry to unlicensed imports, and would not accept the EC's proposed

TABLE 4-2.   EC Import Penetration, Seven Product Categories
of June 3 Proposal, 1977–1981 (percent)

EC Proposal	1977-79	1979-81	1977	1978	1979	1980	1981
6.7	7.6	6.0	8.4	8.3	6.3	5.3	6.4

guarantee of its traditional ⅓ import market share. The US industry (which remained convinced that all the EC producers—including the Dutch and the Germans—would be found subsidized) would not accept the limited product coverage, the flexibility gimmicks, or the liberalization of restraint when capacity use exceeded 87%.

Notwithstanding its clear unacceptability (which had been pointed out several times), Davignon expressed hope that the proposal offered a reasonable basis for settlement. He appealed to Baldrige's desire to "dissipate a possible crisis in a spirit of solidarity." He concluded his letter by attempting to put the burden of reaching an agreement on the US government:

> This agreement is not what our steelmakers would like us to conclude. Likewise I understand that it will not give total satisfaction to the US industry. On our side, it has only been possible by an exceptional effort of the Commission, supported by the member states. I think that it is now to the US authorities to decide upon the position they want to take on this issue.

Finally, Davignon invited Baldrige to come to Brussels "very soon in order to work out a final settlement."

Commerce's first action upon digesting Davignon's letter was to meet at staff levels with EC representatives to clarify and verify the proposal. EC staff confirmed that the 6.7% was not in error; it was the number they had intended to propose and they fully understood its relationship to historical import levels. EC staff said there had been a long battle between the commission and EC producers, and that the commission had beaten the producers back considerably.

An EC diplomat told US officials that the producers believed that neither West Germany nor the Netherlands would be affected by the cases, and that therefore EC export and domestic steel sales could be restructured with no net change in overall EC exports to the United States. Also, the EC producers did not believe that high CVDs would be placed on their exports. Commerce staff, which by now had the draft preliminary CVD results (which would be released in three days), warned that that was a fundamental misconception, and that perhaps the announcement of the preliminary results was needed to convince EC producers of the seriousness of their position. Commerce could not believe that Davignon really thought Commerce could get the US producers to accept the proposal.

On pipe and tube, the EC said there was no argument for its inclusion in the VRA other than "naked economic interest." Commerce staff could not disagree, but noted that whatever moral flavor might be put on the demand, the US industry would not accept an agreement without pipe and tube.

In the staff discussions of the June 3 proposal, some progress was made on a few issues. Commerce agreed to try to convince the US industry to accept the carryover, advance use, and transfer flexibility provisions at low levels, and explained the legal research that had failed to find a way for US enforcement of the restraints. The EC agreed to consult in 1985 on extension to 1987, the first use of provision for further consultations to resolve an issue that could not be resolved at the current time. (The US industry position, based on experience with the vagueness of the enforcement provisions of TPM I and II and somewhat shared by Commerce, was that everything should be nailed down.) The ease with which US and EC negotiators settled this problem made "consultations" an attractive solution for other issues.

Commerce again explained its inability to deny firms their right to file petitions and presented in detail an alternative: the EC could terminate the export restraint agreement if future petitions were filed. The EC had all along been assuming the right to so terminate the agreement—they were searching for stronger immunization. After all, termination of the system would leave the EC subject to a refiling of the very petitions from which it needed the protection of the system. The EC suggested that US producers submit letters promising not to file petitions for the duration of the program. Commerce noted that such commitments would probably not be enforceable under US law, but that the idea of a publicly stated intention not to file petitions was attractive.

Commerce now set strategy for the last few days remaining until June 10. Once again, the EC proposal could not be given to the US industry because it would make them furious. Therefore, Commerce prepared a draft agreement that might be acceptable to both sides. The draft was as accommodating of EC concerns as Commerce could be and still have some chance of success. After one more round of discussions with the US industry, Commerce would redraft it and present it to Davignon as the last opportunity to reach agreement before the CVD preliminary determinations.

The further round of discussions with the US industry bore some fruit. The producers' representatives opposed allowing the EC to terminate the system if a single petition was filed; each firm feared that a producer of an unimportant product would file a petition, the EC would kill the system, and the industry would have to file petitions and wade through more investigations to regain relief. They agreed to pledge not to file petitions for the duration of any arrangement. General agreement was reached on Commerce-proposed language to allow limited increased EC exports if supplies of a given product were short.

Overall, however, the US industry seemed to oppose the agreement as drafted (even though in its most recent draft Commerce had deleted the "flexibility"). The absence of pipe and tube coverage and the lack of US enforcement were critical. Ominously, representatives of the US firms began bidding up the product coverage requirements: two firms now wanted rails included, one said *all* major carbon steel products should be covered, most wanted specialty steel, another wanted tin plate, another alloy bar, wire rod, and wire and wire products.

The proposed restraint level was discussed with each firm individually to avoid potential antitrust problems. Commerce now proposed 4.33% for all steel mill products, a return to the original government proposal. The 4.33% on all products translated into 5.2% for the seven, well below the EC's proposed 6.7%. While several firms rejected 4.33% in preference to 3.95%, Commerce officials felt fairly certain that if most other major problems were eliminated, 4.33% would be acceptable.

Upon completion of the talks with the steel company officials, Commerce constructed its eleventh-hour offer for the EC, a proposal that had not been approved by the US industry but that Commerce officials believed would be acceptable to them on the whole if pipe and tube were added. In recognition of the EC's claim that addition of pipe and tube was impossible, it was left out of the draft transmitted from Baldrige to Davignon on June 7, but in an accompanying letter Baldrige warned that he believed no agreement was possible absent a pipe and tube agreement.

Baldrige explained his position on the level proposed by Davignon:

> The 4.33 percent (proposed here) was based on reducing EC import levels from the recent past, much as you did in your proposal. [Recall that 3.95 had been proposed originally because it

was justifiable; here the pretense of an objectively derived number was abandoned.] US producers could not accept the figures you proposed, given their belief that their cases will result in a far larger reduction. The market share we propose, while substantially lower than one you and your producers would like, is above the level of 1980 imports.

Baldrige tried to meet EC goals within the latitude available to him. Rather than propose that the system endure to 1987, he proposed consultations in 1985 to consider extension. Rather than simply affirm US government inability to refuse petitions, he offered written assurances from the petitioners that they would not file cases and a recognition of the EC right to terminate the *entire* VRA if any petition was filed on a covered product. Rather than reject US enforcement, he promised to continue to search for an appropriate legal means, and pledged that if one was found he would bar entries of unlicensed exports. He did not yield on the flexibility gimmicks, however—carryover, advanced use, and transfer were deleted from this draft, on the theory that the other flexibility in the mechanism (reconciliation of forecast to actual consumption quarterly adjustments, and the short-supply provision) would be sufficient.

As for product coverage, Baldrige communicated bad news: he now believed that three additional products had to be covered: cold-finished carbon bar, cold-finished alloy bar, and tin plate. Baldrige told Davignon that he had scheduled a meeting with the CEOs for the afternoon of June 8 and that he hoped to hear from him before then.

We must note here the importance of time factors. With the deadline for preliminary CVD determinations so near and so many interested parties, the Commerce Department's attempts to frame an agreement acceptable to all were in part educated guesses. Neither the EC nor the various components of the US industry addressed themselves *in detail* to the various proposals advanced until, on several occasions, it became more likely that an agreement would be reached. Then, details that had been overlooked previously were noticed and changes demanded. In part, the late focus on detail was the result of a broadening of the circle of participants at this stage to include technical experts of the steel producers, such as marketing people and additional lawyers. Baldridge's last-minute addition of tin plate and cold-finished bars is attributable to a late plea from US producers and Baldrige's guess that those producers would not agree to withdraw their petitions without the added coverage. Ultimately, Baldrige's educated guess on cold-finished bars turned out to be wrong, and agreement was reached without that coverage—much to the consternation of specialized US cold-finished bar producers.

Davignon's response was punctual, if not encouraging. The EC Council of Ministers, the highest policymaking body of the community, had met to consider Baldrige's proposal. It rejected it for four main reasons: pipe and tube remained impossible to include; US government cooperation was needed for the efficient functioning of the agreement—that is, the EC required US enforcement; a stronger guarantee that no petitions would be filed during the agreement was needed; and the proposed reduction in quantities was too great, *and* the reduction was sought from the EC only. The EC indicated a willingness to continue talking, but further talks had to wait until after the announcement of the preliminary CVD results.

5 Preliminary CVD Determinations and Subsequent Negotiations: June and July 1982

THE PRELIMINARY CVD DETERMINATIONS: JUNE 10, 1982

On June 11 Commerce announced the results of five months of investigation, confirming many of the EC's worst fears while seriously limiting the US industry's negotiating possibilities. Commerce had basically adopted the position that any government provision of goods or services on noncommercial terms that was targeted to a particular firm or industry constituted a countervailable subsidy.

The targetting criterion forced Commerce to refuse the only US industry argument that could have resulted in substantial countervailing duties against West Germany: that acknowledged subsidies to the German coal industry constituted countervailable subsidies to the German steel industry. Neither did it accept other contentions that would affect German and Dutch steelmakers, such as that investment tax credits available to many industries are countervailable subsidies. Both those steel industries were found to be essentially unsubsidized (subsidy rates of less than one percent).

The flip side of those determinations was in some cases staggering levels of subsidies found in other EC countries. British Steel, the recipient of government equity infusions and low-interest loans, was found to be subsidized by (and thereby subject to countervailing duties of) 40%. The major French, Italian, and Belgian producers were found to have received similar subsidies in amounts ranging from 20 to 30%.

Detailed consideration of the subsidy issues addressed in the investigations is beyond the scope of this volume. The lawyers for both the EC and the US producers attacked the results, but Commerce was confident that even if some details of its determination were found wanting, the broad outlines of its decision would stand up (in part because it was no secret that billions of government dollars had gone into the Italian, British, French, and Belgian steel industries, while little or none had gone to the Dutch or German firms).

Baldrige put the dispute in perspective and left the door open to continued negotiations in his public statement accompanying the preliminary results. At the same time, he left no doubt that he was prepared to carry the investigations through, come what may. He first noted his sympathy for the distressed US industry and his determination to eliminate the unfair advantage subsidies gave certain foreign producers. He warned the US industry that eliminating unfair trade would not solve all of its problems and, in a message to both the US industry and the EC, said that "the US government will take no action against fairly traded steel imports." He then turned to the EC and the cases:

> For several foreign producers, their employees, and their governments, today's action could be a serious setback, despite the fact that export of the products under investigation accounts for only about 3 percent of the EC's steel production. The European steel industry has been in crisis for almost five years, and the possible loss to the more heavily subsidized European steel producers of much of their US exports is one more cross for that industry to bear.
>
> I regret it has become necessary for the US government to take the steps announced today, but remind all that these cases were provoked by the combination of the 1981 surge in exports to the United States, and continuing, and in some cases growing, government financial support.
>
> From the time these cases were filed in January, I have stated that I am fully prepared to carry these cases to conclusion, and I remain partial to clearing the air by doing so unless an equitable settlement can be reached.

With the June 10 determinations, the Commerce Department eliminated much of the uncertainty that had plagued the early negotiations. While both sides would continue to fight the technical subsidy issues, it was now clear that the Dutch and Germans would likely escape any serious consequences from the cases (while the dumping cases remained to be completed, steelmakers in those two countries were confident they had not been dumping—the dollar was strong enough to protect them). The US industry would therefore look more favorably at a negotiated settlement in order to prevent the Germans from taking up the slack in the US market from the Belgians, British, and others. The Germans were thought to have both capacity and the incentive to do so, particularly if the subsidized EC producers redirected their former US sales to West Germany. At the same time, German steelmakers, who had been concerned that the US government would be unfair in its subsidy rulings, had reason to further stiffen their resistance to compromise. In contrast, the British, French, Belgians, and Italians were forced to confront the certainty that the Reagan administration would enforce its countervailing duty law and close its steel market to those already beleaguered steel industries. To those countries, the need for settlement heightened.

The official EC response to CVD preliminaries came in two stages. The first was protest combined with a "grave concern":

The EC has expressed its grave concern to the Administration that this decision will disrupt traditional trade flows. Steel exports of certain EC Member States will be virtually eliminated from the US market. . . . other EC Member States . . . will suffer from a backlash on the domestic EC market.

The Department of Commerce decision will certainly undermine the Community's restructuring policies. It will seriously aggravate the problems of excess capacity in Europe, lead to a lowering of prices on the internal EC market, and growing unemployment.

The US decision, which reflects an extraordinarily rigorous interpretation of US laws, appears to contain a far-reaching interpretation of the GATT Subsidies Code that the US Government doesn't seem ready to apply to certain US programs on which conclusions have been reached in the GATT.*

Over the past few weeks, the Commission has spared no effort to avoid the present situation. The Community thus regrets the Department of Commerce decision even more so because it should have been unnecessary.

Eurofer, the EC steel producers' association, quickly and interestingly (given assumed German and Dutch opposition) expressed its unanimous belief that the EC should continue toward a negotiated settlement.

---

*This referred to the US DISC export subsidy program, with considerable justification.

The second stage of the EC response came after the US government imposed sanctions on EC firms building equipment under US license for the Soviet gas pipeline to Europe. On both sides of the Atlantic, concern over the fundamental strength of the alliance was being expressed. On June 22 the EC Council issued a call for reconstruction, noting that the steel and pipeline disputes "should be seen against a background of escalating trade disputes between the United States and the Community, not just in relation to steel, but also to agriculture, export credits and textiles." It regretted the "unilateral nature of the US response to these problems" (steel and pipeline), and considered that "action is needed at the highest levels to find solutions through constructive decisions."

The United States recognized the need to ameliorate the dissension in the alliance caused by the fights over steel, agriculture, and East–West trade. A senior interdepartmental review group of relevant Cabinet officers was formed and asked by the president to find a way to improve US–EC relations. Baldrige simultaneously moved to reopen negotiations on steel.

COMMERCE AND EC MOVE, BUT US INDUSTRY REMAINS ALOOF: EARLY JULY 1982

On July 2 Baldrige launched another attempt to secure an agreement. At a breakfast meeting with David Roderick, CEO of US Steel and president of AISI, Baldrige said he intended to begin discussions with the EC once again. He promised to insist that pipe and tube be covered by the agreement, but warned that coverage in the same manner as other products was unlikely. Roderick had little to add to previous industry positions. Baldrige and Olmer then traveled to Brussels to talk with Davignon.

Baldrige made two major concessions to the EC at this stage. The first was that pipe and tube could be handled differently than other products. The second was to agree to refuse entry to unlicensed EC exports. Commerce staff, in consultation with other relevant agencies, had still not found a legal means to do so, but were willing to "do something" even if it required a stretching of existing laws, which would almost certainly be overturned in a court challenge. Ultimately, this concession meant that legislation would be needed.

Armed with a new indication of US flexibility on pipe and tube, the EC went back to its pipe and tube producers and found they were willing to consider some type of agreement so long as it was separate from an agreement covering ECSC products. The producers also said they could not enter into substantive talks until mid-August, ostensibly because they could not meet together before then.

The period of reflection since the June 10 CVD preliminary determinations had revealed additional difficulties to both sides. EC producers were concerned about the effective date of a VRA; given the proposed level and actual exports to date in 1982, entry into effect in the third quarter of 1982 would leave very little allowable exports for the rest of the year. In addition, they were not satisfied with US producers' assurances that no cases would be filed; they wanted at least some government guarantee. Of course, EC producers felt the government's June 7 proposed level was draconian, but they were not prepared to discuss levels until a product coverage decision had been reached. To EC producers, product coverage and levels were linked: the broader the product coverage, the less the cutback in the levels.

Product coverage remained intractable even though pipe and tube had been set aside for the moment. The EC Commission would not (it said it could not) cover specialty steel, although it had previously offered to cover some specialty products. United States negotiators, recognizing the EC's inability to control small specialty producers, proposed that the EC cover only those products made by larger firms over which it had more control: stainless steel sheet, strip, and plate. This idea seemed to offer an opportunity for compromise. The commission maintained that EC producers strenuously rejected all other increased product coverage that the US industry demanded.

After a week or more of intense senior-staff-level discussions in which no meaningful progress was made, Baldrige approached Davignon directly as he had done in early June and asked for a decision: What was the EC willing to offer at this stage? Without some movement from the EC, further staff discussions would be meaningless. Davignon responded with major concessions, apparently taken on his own without prior agreement from EC producers. Stainless steel sheet, strip, plate, rail, tinplate, and hot-rolled carbon bars would be added, quarterly consultations would be instituted with respect to other bar products, and the level for the initial seven products would be reduced from 6.7% to 5.67%—subsequently changed to 5.9%. This offer was conveyed to the US steel producers, who not only were not impressed, but chose this moment to add another demand for product coverage: semifinished products. Baldrige passed the rejection back to Davignon, who was very disappointed to hear of the out-of-hand rejection, with the difficult if not impossible (and to Davignon totally unjustified) demand for semifinished products added.

Davignon's strategy—taking responsibility for concessions himself—had failed. It seemed to the EC Commission, which now had angered its steel producers by the concessions on product coverage and export levels to no end, that there was no possibility of reaching a reasonable agreement. Davignon therefore returned to an old idea: the US government should reach a reasonable settlement without submitting it to the veto of the US industry. He proposed that agreements be made through the suspension agreement provisions of the US countervailing duty law (which permitted such a deal under certain limited conditions). The US industry's greed on product coverage—first on pipe and tube, and finally on semifinished products—and general unwillingness to wrestle with Davignon's latest proposal put Baldrige in a mood to look more favorably on Davignon's latest tack.

SUSPENSION AGREEMENTS: LATE JULY 1982

Davignon's suggestion for a change of course came late on July 21, only three days before the last possible date for reaching such an agreement and a few days after initial feelers from British Steel had been received on an unofficial and informal basis. It is not clear whether Davignon knew of the British Steel contact with the Commerce Department and had moved quickly to keep the commission in charge of events, or whether he independently thought the change of course appropriate. In any case, it appeared that British Steel had grown impatient with continued failure of the US–EC negotiations and was contemplating striking out on its own if need be, forced by the approaching statutory deadline for suspension agreements.

Section 704(c) of the Tariff Act of 1930, as amended in 1979, stood at the center of the new controversy. That section allows the Commerce Department to suspend a countervailing duty investigation pursuant to a quantitative restraint agreement with a foreign government. The term "suspension agreement" is applied because the CVD investigation lies in suspension, ready to spring to life if the agreement is violated. The law provides three criteria such an agreement must meet: it must completely eliminate the injurious effects of imports of the merchandise under investigation; it must be more beneficial to the domestic industry than completion of the investigation; and it must be in the public interest.

The law states that an agreement must be initialed at least 30 days prior to the final CVD determination (in this case, by July 24, 30 days prior to the August 24 deadline). The domestic industry then has 30 days to comment on the proposal, following which the agreement may be signed by the Commerce Department and the foreign government, even over US industry objections.

The US industry does have some recourse if it is not satisfied with a quantitative suspension agreement. It may request that the ITC review whether injury has been completely eliminated; or it can go to court to challenge an agreement on a number of bases, including that the agreement is not in the public interest or that it is not more beneficial to the industry than completion of the investigations. Because of the high subsidies found in the preliminary determinations, it would be very difficult for Commerce to argue that any nondraconian VRA was better for the industry than continuing the investigation if the industry itself vehemently felt otherwise.

In unofficial discussions with low-level Commerce staff prior to Davignon's shift of course, British Steel attempted to deal with the "more beneficial" question realistically. While under the statute only products subject to investigation could be covered in a suspension agreement, British Steel contemplated annexing a voluntary undertaking to its suspension agreement, limiting the export of several products not subject to investigations. The EC, in the proposals it presented officially on July 22, chose to ignore the "more beneficial" problem.

On the morning of July 22, the EC issued a press release announcing its decision to propose suspension agreements. Davignon sought to portray the proposal as an opportunity for the US government to decide on its own, without steel company veto power, whether it wanted an agreement. He told the press

> this confronts the American Government with its responsibilities since it alone must reach a decision on the basis of its own legislation. The political significance of today's decision is that henceforth things will be clear. Either escalation is avoided, something which the Community wishes to see, or the American Government rejects our offer, our last, made in good faith and which goes farther than the narrow framework of steel. In this latter case, each of you can imagine what the consequences of a negative response from the US would be.

The impatience of some of the member states, particularly the UK, had become public knowledge, and Davignon stressed that the community was maintaining solidarity.

The community proposed that suspension agreements cover the major products under investigation from the high-subsidy countries only—Belgium, France, Italy, and the UK. The FRG, Netherlands, and Luxembourg were "prepared not to take advantage of the export limitations" on the other countries. The EC required that the United States accept the proposals with respect to all products and countries, or none of them.

The proposed agreements were virtually identical in mechanism to the latest draft VRA and included some of the same technical points to which the United States had consistently objected—for example, the short-supply provision based solely on aggregate US steel capacity utilization. However, those points were overshadowed by the proposed levels: a mere 10% reduction in market share by product from 1981, the year whose imports had triggered the petitions. The 10 percent reduction from 1981 was unacceptable in large part because the combined market share of all products for which suspension agreements were proposed had risen from 2.74% in 1979–80 to 3.41% in 1981, an increase of 25%. The ten percent reduction would leave the EC with a larger market share in these most-heavily subsidized products than before the 1981 dispute began. Given the legal requirement that injury be eliminated and the "more beneficial" requirement, this was clearly not enough of a reduction. The Commerce Department knew immediately upon receiving the proposal that it could not be accepted.

The unrealistic proposals were disappointing to US officials, for the 704(c) provision would have allowed for US enforcement. Failure to use the 704(c) opportunity would leave the fate of an agreement entirely to US petitioners. Commerce officials were not certain whether the EC proposal represented a serious attempt at suspension agreements without US industry consent (the inclusion of unrealistic levels and unacceptable mechanisms on a last-minute, take-it-or-leave-it basis suggested not), or an attempt to assert EC control over member states, or simply another attempt to put the onus of failure on the United States.

The question facing Commerce officials was not whether to reject the proposal, but when and how. The UK, knowing that the United States would have to reject the EC proposals, was pressing for an immediate rejection without a counteroffer so that it would have time to conduct bilateral negotiations with the Commerce Department. The UK recognized that the EC proposals were unacceptable, and believed its ideas, which were unacceptable to the EC, would make agreement possible.

United States officials were of two minds on the British desire for a bilateral deal. On the one hand, they were sympathetic. British Steel's American CEO, Ian MacGregor (who was friendly with Baldrige), had taken drastic steps to put the firm back on a competitive footing. He had closed entire divisions, slashed employment and capacity, and sought to end subsidization as soon as possible. On the other hand, US officials were reluctant to be active participants in settlement discussions without EC.

Not certain of EC attitudes toward bilateral agreements and wishing to keep the possibility of separate bilateral suspension agreements alive, Baldrige rejected the EC's proposal only minutes after it was formally received at 5:30 P.M. on July 22 (hours after the contents had been informally conveyed). Baldrige said in rejecting the proposals:

> unfortunately these proposals are not legally acceptable. US law requires that any quantitative restraint agreement must completely eliminate the injurious effect of subsidized exports . . . the 10 percent reduction offered today . . . was simply not sufficient to remove injury.

Baldrige also noted that even if he could have accepted the proposals, trade tension would not have been eliminated because the steel antidumping cases would remain. A Commerce counterproposal, with levels based on the likely effects of the preliminary countervailing duty decisions, was cabled at once to Brussels, with no real hope that an agreement could be reached in the two days remaining.

The next day (Saturday, July 23) the British under secretary of state for industry came to Washington for exploratory talks on a separate bilateral agreement with the United States. The UK did not have permission from the EC to negotiate separately with the United States but, because only two days remained until the deadline and many complex issues remained to be resolved, felt compelled to talk nevertheless. The United States was in a similarly difficult position, not wishing to offend the EC but not wanting a possible bilateral agreement with the UK to fail because of lack of time. Thus, "nonmeetings" were held on a technical level separately with the EC, the British, and the French (further complicated by the inability of the ranking Commerce staff person to talk with the British, because of prior representation of British Steel), while higher officials on both sides of the Atlantic pondered whether bilateral agreements should be allowed.

The issue was resolved on July 24 when the EC Council of Ministers, after a ten-hour meeting, formally rejected the Commerce counterproposal, but gave the commission an exclusive mandate to negotiate with the US government on steel, effectively barring bilateral suspension agreements. Further negotiations would now, by virtue of the passing of the July 24 deadline, depend on the assent of the US steel industry.

6 Commerce Takes a Stand: The August 5 Agreement

With the possibility of an anarchic rush of member states to the United States to cut their own deals over (for now, at least), the United States and EC redoubled their efforts to reach an agreement. As a result of the near-collapse of EC control and a heightening of US–EC tension, primarily because of US sanctions related to the Soviet gas pipeline, the US government made a fundamental change of strategy. Baldrige now sought an agreement with the EC that he could recommend to the US industry. The "success" of a steel agreement in some sense now became less important to the two governments than the need to show that they could reach agreement among themselves.

Immediately after the EC Council decision on July 24, US negotiators returned to Brussels, and then EC negotiators followed them back to Washington. By August 5 Baldrige and Davignon were satisfied with the draft agreement and announced their acceptance of it. Both agreed to recommend it to the remaining parties (the EC Council and the EC and US steel industries). Within hours of release of the text, several US petitioners rejected the agreement. In fact, the US specialty steel industry rejected the agreement publicly while Commerce officials were still presenting it to steel company representatives.

This rejection was not surprising, for there was little new in the August 5 agreement. The proposed levels were too high, pipe and tube coverage inadequate, and product coverage too selective. Any one of these shortcomings was enough to cause rejections, and there were other problem areas as well.

Product coverage remained at the original seven products plus tin plate, rails, and stainless steel sheet, strip, and plate. Pipe and tube was dealt with in a draft side-letter, but no detailed export restraint was specified:

> Given the magnitude of trade in the pipe and tube sector both parties recognize that it is necessary to ensure that trade distortions in that sector do not arise which would undermine this arrangement and create problems in that sector. To this end, discussions should therefore rapidly be engaged on this matter and it is on this assumption that the Department of Commerce agrees to this arrangement. The nature and the objective of these discussions are not the same as for the arrangement, nor will the legal means by which the results will be attained. Both parties undertake to use their best endeavours to resolve this matter [pipe and tube] by 15 September.

The restraint levels were set in the aggregate at about 1979–1981 levels, with individual products above and below that level. Because the exact manner in which quotas are set is usually so mysterious and of much interest, it is worth presenting a detailed description of how the August 5 numbers—which ultimately became the numbers in the final arrangement—were reached.

The numbers began with the EC proposal, first suggested in the July 22 suspension agreements for a few products most affected by the CVD cases, that the import market share of products subject to investigation be reduced by 10% from the 1981 level, and that the market share of products covered by the arrangement but not subject to investigation be set at the 1981 level with no reduction. However, the Germans and Dutch objected to a 10% reduction in cold-rolled sheet exports (which they dominated), and convinced the EC that the reduction should be less (because, even though there were investigations of cold-rolled sheet, the Germans and Dutch were convinced they could win the cases). The EC also had made a mistake in calculating historical US consumption of coated sheet, resulting in a proposed market share higher than actual 1981; the mistake in the consumption figure was corrected but the higher market share still stood. The mistake probably stood because EC negotiators stated that under no circumstances could they agree to an aggregate market share of covered products of less than 5.76% (notwithstanding warnings from US negotiators that the petitioners would not withdraw their petitions at that level), and every last ton was needed to meet that figure. Besides, 1981 had been a very bad year for EC coated sheet exports.

Regardless of how arbitrary the 5.76% figure was, EC negotiators would not move. The general scheme—10% off for covered products, 1981 levels for uncovered products, special deals for cold-rolled sheet and coated sheet—resulted in 6,000 tons less than needed to meet the 5.76%. United States and EC negotiators thus distributed the 6,000 tons into product categories where it would cause the least trouble (tinplate, hot-rolled sheet, and structurals). The 6,000 tons made very little difference in market shares: for example, the EC share for structurals was boosted from 10.87 to 10.90%.

The proposed level for all covered products was about 5.75% of US apparent consumption of those products, down 8% from 1981's 6.25% and almost identical to the 1979–1981 level of 5.78%. Baldrige had in effect acceded to the EC's wish to maintain its traditional market share. Rejection by US industry was certain; it would not accept the import levels that had provoked it to file petitions in both 1980 and 1982, at a cost of terminating the AD and CVD cases that held promise of at least some initial reduction in imports (until the Germans could move into French, British, and Italian markets) and a certain measure of revenge on the subsidized EC producers.

Other aspects of the August 5 arrangement were not immediately seized upon as reasons for rejection but would become important later on. The VRA was to begin on October 1, assuming petition withdrawal by September 15, and transition period exports were to be limited in a vague manner. Limited advance use, carryover, and transfer were allowed, as were increased exports of specific products when "the US in consultation with the ECSC determines that because of abnormal supply or demand factors, the American steel industry will be unable to meet demand in the USA for a particular product." The United States agreed to prohibit entry of unlicensed merchandise, but still could not specify how it would do so.

The immunization issue was settled to the satisfaction of both sides by adding an ambiguous clause to previous language. The EC retained its right to terminate the export restraint on any or all products if a petition was filed on any covered product. In addition, if a petition were to be filed on any *other* steel product such that the "objective of the arrangement" would be threatened, "then the ECSC and the US, before taking any other measure, shall consult to consider appropriate remedial measures." United States negotiators believed this to mean that the industry was free to file petitions against uncovered products, while the EC took it to mean it had the right to terminate the system if a petition was filed against *any* steel product, covered or not.

The first to reject the August 5 agreement was the specialty steel industry. It objected to the partial coverage of specialty steel and to the proposed level of 4.08% of US consumption. The 10% reduction from 1981 import levels resulted in a level far above recent years (see Table 6-1).

The major carbon steel producers spent a bit more time considering the proposal, but virtually all rejected the agreement within 24 hours. David Roderick, chairman of US Steel, reluctantly rejected the agreement, noting only that the levels were too high. National Steel's chairman, Howard Love, in a letter to Baldrige also focused his criticism on the proposed levels:

> The proposed arrangement would establish import levels for the Europeans which are only slightly lower than 1981 figures and virtually unchanged from the three-year average for 1979 through 1981. It was the high level of imports in 1981 that caused us to take action initially. We also believe that the Europeans must, in some way, be held accountable for their violations of US trade law in the first half of 1982.

TABLE 6-1. **EC Market Share of Stainless Steel Sheet, Strip, and Plate, 1977–1981 (percent)**

1977	1978	1979	1980	1981	1979–1981
2.3	2.5	1.4	1.2	4.5	2.4

Bethlehem Steel provided Commerce with a more detailed list of criticism of the agreement. A major structural producer, it found unacceptable the 10.9% market share, which was above the 1979–1981 average of 10.7%. Its general counsel was not satisfied that the government could deny entry to unlicensed merchandise and therefore had little confidence in its enforceability. It also pointed out several product coverage gaps that had not been previously raised as requirements, including alloy products, black plate and tin-free steel, and wire and wire products. All of these criticisms would be repeated and expanded upon when the industry got serious about negotiating an agreement—after the preliminary dumping duties were announced on August 10 and the final CVD determinations were announced August 25.

Without US industry agreement, the August 5 arrangement failed to resolve the steel dispute. It did, however, ease the tension between the US government and the EC Commission, which at the time was the most important goal.

7 More Commerce Determinations

THE PRELIMINARY ANTIDUMPING DETERMINATIONS: AUGUST 9, 1982

The inescapable message of the announcement of preliminary anti-dumping margins, which in effect was precisely parallel to the preliminary CVD determination, was that most EC steel producers would not be found to be dumping in the United States.

A first examination of the dumping determinations might not reveal the magnitude of the damage to the US producers' negotiating position. Dumping margins as high as 41% (Teksid of Italy) were tentatively imposed and Thyssen, the largest German producer, was found to have dumped steel sheet at a rate of 19%. These margins were misleading, however, because those respondents submitted information too late to be used in the preliminary determinations and Commerce had been forced to use the petitioners' claims instead. With the exception of British Steel (with high costs and a currency relatively strong because of North Sea oil) and Italsider (Italy), in no case was an EC producer that provided adequate information found to be dumping significantly.

The clearest decisions concerned some major German firms, the Luxembourg and Dutch producers, and Sidmar, the modern coastal Belgian plant, all of whom were found not to be dumping. Usinor, the only French company to supply good data, was found to be dumping at a 1 to 3% rate, and Cockerill-Sambre, an inefficient Belgian producer, was found to be dumping at only 1, 5, and 11% on three products.

We must here caution the reader about the proper interpretation of dumping margins. The calculation of dumping margins involves a mass of complex manipulations, including properly matching home market to US sales, adjusting for cost differences that would justify price differences, and choosing the correct exchange rate. The basis for fair value differed case-by-case, but in most of these cases home market prices had been below full accounting cost for some time; consequently "constructed value"—fully absorbed average production cost plus a statutory minimum of 10% for selling, general and administrative expenses, plus a statutory minimum 8% for profit—had to be relied upon. These minimum figures may bear little resemblance to reality in any industry, and to base the profit figure on cost is without financial or economic merit. Finally, subsidies were not included in the dumping calculation (to avoid double-counting in parallel AD and CVD cases).*

It now seemed more likely that West Germany, the Netherlands, and Luxembourg would be unscathed by the unfair trade cases and free to expand US exports without fear of AD or CVD cases, and that only countervailing duties would restrain the rest of the EC. A negotiated settlement with broader product and country coverage looked more attractive to the US industry.

One aspect of the decision shocked the Europeans, demonstrating to them once again that Commerce officials would have to enforce the law, regardless of diplomatic backlash. "Critical circumstances" were declared in four cases. Importers of products affected by these declarations would be required to pay duties on not only all future imports, but also on all imports received in the previous 90 days because "there have been massive imports of the subject merchandise over a relatively short period of time and the importer knew, or should have known, that the merchandise was being dumped or there is a history of dumping." (This provision had been added to the law in 1979 as a compromise over the AD and CVD laws' prospective nature.) The provision had never been used before and provoked the usual barrage of EC criticism, but Commerce made clear its willingness to enforce the law.

THE FINAL CVD DETERMINATIONS: AUGUST 24, 1982

The final CVD rulings did nothing to improve the apparent bargaining position of the US steel industry, as nearly all of the rates declined from the

---

*A fundamental rethinking of the antidumping law is in order.

preliminary levels. Table 7-1 shows the decline in calculated subsidy rates for four of the largest, most subsidized EC producers. Most important, Germany, the Netherlands, and Luxembourg were confirmed to be unsubsidized as were nearly all the small private British producers, one large Belgian producer, and even the state-owned South African producer. A rough calculation showed that of the 1981 US imports of EC products still covered by CVD investigations (after the ITC had eliminated many cases in the preliminary injury phase), only 36% would be subject to countervailing duties of 5% or more. This amounted to only about 28% of 1981 US imports of nonpipe and tube EC steel. The CVD cases would cripple a few major EC exporters; they would not greatly reduce steel imports in total.

TABLE 7-1.   Comparison of Preliminary to Final CVD Rate, Selected Companies (percent)

Firm	Country	Preliminary Rate	Final Rate
Cockerill-Sambre	Belgium	21	13
Sacilor	France	30	14-20
Usinor	France	20	11-18
British Steel	U.K.	40	20

The reduction in subsidy rates from the preliminary levels was primarily due to reversal of policy on two important technical points. First, Commerce in the preliminary determination had allocated subsidies given to cover operating losses over the useful life of steel mill equipment. For the final determination Commerce decided that such subsidies are more accurately allocated to the year in which the loss was incurred. Thus, all subsidies prior to 1981 that were considered to be covering losses (and all cash infusions were assumed to go first to cover losses) were ignored in calculating subsidies. Second, Commerce in the preliminary determination had used estimates of the cost-of-capital as a discount rate for the allocation of the money over time, but in the final determinations decided that it was not fair to impute to a company a discount rate higher than the risk-free rate in that country. Because the amounts of money involved in subsidies for "loss coverage" and in grants allocated over time were so large in these 1982 steel investigations, the shifts in subsidy rates were large.

Even though Commerce officials made these technical decisions on the basis of the law and the facts as they understood them (and after intense review and consultations with experts), wags in and out of the department saw the decisions as part of a "subsidy reduction program." The supposed aim of the program was to ease the burden of the projected duties on EC firms and/or to help convince the US industry to be more flexible in its bargaining. The Commerce Department has since reexamined its policies on these two technical issues and essentially returned to the positions taken in the 1982 preliminary determinations.

After the August 24 final determinations, only about two months remained until the CVD investigations would be finished and the duties definitively imposed. In those two months, the ITC had to complete its final CVD injury determinations, and Commerce would continue toward its final AD determinations (due in late December). While it is possible to revoke countervailing duties after they are imposed under certain conditions (which here meant the consent of the petitioners), the legal process was cumbersome and untried. Commerce's legal counsel estimated that procedural requirements of a termination would take at least six months. Therefore, the ITC final CVD decision was a very important deadline, and became the goal for the negotiators. If this new deadline was missed, all believed, there was no serious chance of reaching a negotiated settlement (in part because millions of dollars would have been due the US government in duties).

8 The US Industry Responds: August 25, 1982

There was very little official EC response to the preliminary anti-dumping determinations and, as expected, continued strong protest at the subsidy principles underlying the final CVD rulings. In contrast, the EC response to the August 5 agreement and its immediate rejection was strongly positive. The EC lauded US–EC cooperation and publicly treated the issue as resolved: according to the EC, it was only up to the US government to deliver its side of the bargain. Baldrige tried his best to convince the US industry to go along, but the agreement had not been designed with that in mind. Nevertheless, the August 5 arrangement did, along with the preliminary AD decisions, focus the industry's attention on the need to seriously address negotiations.

The day after the CVD determinations were announced, David Roderick, chairman of US Steel and AISI, presented a detailed proposal for alterations in the August 5 agreement that would make it acceptable to the industry (for which Roderick said he was speaking). His proposal was not incorporated in a document. Instead, he presented it orally to Under Secretary Olmer and Deputy Assistant Secretary Horlick, leaving the Commerce Department with only Horlick's notes and the pair's memories upon which to rely. Commerce officials later learned that Roderick's presentation mixed two distinct proposals, one of them US Steel's and another from other industry members. After staff found serious inconsistencies in Roderick's presentation, lower-level US Steel officials explained the two proposals and claimed that Roderick had convinced the rest of the industry to go along with US Steel's plan (which was somewhat less demanding than that of other firms). Roderick, like Davignon earlier, was taking a gamble that he could deliver his constituents. Many of the tougher demands alluded to by Roderick but then passed over were to be revived a month later, at which time they would nearly kill any chance of reaching an agreement.

The US industry proposal had five major elements: reducing permitted export levels; broadening product coverage; providing for consultations to keep exports of products not restrained by export licensing from growing; reducing the right of the EC to terminate the system if unfair trade petitions were filed; and a side-agreement to keep the EC share of the US pipe and tube market at 5.9%, the 1979–1981 average (compared to 1981's 10.9% share). Roderick granted one concession: the demand that semifinished steel be covered was dropped, a product-coverage gap that Roderick said would lead to major US plant closings.

As Roderick explained the proposal, the extension of product coverage and the change in export levels were linked. Additional products, all of which had been mentioned previously by one or more US producers, were rolled into the product categories of the August 5 arrangement, and a specific market share proposed for each new, broader category. Alloy products were added to the hot- and cold-rolled sheet, plate, wire rod, and coated sheet categories; terne plate and sheet added to the coated sheet category; and black plate and tin-free steel added to the tin plate category. One new category, sheet piling, was to be added at the 1981 level, and reductions with no change in product coverage were requested for hot-rolled carbon bar and rail.

Comparison of the US industry's proposed levels to the August 5 arrangement levels is complicated because of the differing product categories. US Steel presented the plan as simply adding previously uncovered products at their 1981 level to existing categories, in essence accepting the August 5 levels *if* product coverage was broadened. However, a close analysis of the numbers revealed that those proposed by Roderick were not exactly those obtained by following Roderick's rule (that is, adding new products' 1981 levels to existing categories). Table 8-1 summarizes the industry's proposals and shows how Roderick's levels are not strictly derivable from the August 5 arrangement.

The questions this proposal raised in the minds of Commerce staff—which went beyond nonderivability of the levels and led Commerce to double-check the proposal with US Steel officials—are evident from a close examination of the table. Some of the proposed changes were trivial (for example, coated sheet); most changes were small; and some *increases* in import levels apparently were contemplated. Overall, a reduction in levels (aside from product coverage additions) of only 3.5% was requested. How-

TABLE 8-1. **Comparison of August 25 US Industry Proposal to August 5 Arrangement (percent)**

Product Category	Change Requested	US Industry Proposed Level[a]	Adjusted Arrangement Level[b]	Difference
Hot-rolled Sheet	Add Alloy	7.0	6.82	3
Cold-rolled Sheet	Add Alloy	4.9	5.11	-4
Plate	Add Alloy	5.06	5.36	-6
Structurals	Add Alloy	9.5	9.91	-4
Wire Rod	Add Alloy	4.29	4.61	-7
Tin Plate	Add Black plate and tin-free steel	2.2	2.04	8
Coated Sheet	Add terne plate and sheet	3.32	3.27	1.5
Sheet Piling	New Product	21.85	—	—
Rail	Reduce Level	7.3	8.9	-18
Hot-rolled Carbon Bar	Reduce Level	2.0	3.01	-34
Stainless Steel Sheet, Strip, and Plate	Reduce Level	2.4	4.08	-41
Totals[c]		5.22	5.41	-3.5

[a]For the expanded product categories.

[b]For example, in hot-rolled sheet, the August 5 arrangement market share of *carbon* hot-rolled sheet of 7.43% times 1981 US apparent consumption of that product of 14,640,403 gives hypothetical carbon hot-rolled sheet imports of 1,087,782 tons. Add actual 1981 US imports of 122 tons of hot-rolled *alloy* sheet from the EC to get hypothetical imports of the new category of 1,087,904. This total is then divided by 1981 US apparent consumption of hot-rolled *carbon and alloy* sheet and strip of 15,946,834 to arrive at the adjusted arrangement level of 6.82%.

[c]Using 1981 as the base year.

ever, quota construction is not a logical endeavor, and the proposal was accepted—after reassurance—as genuine.

The requests for coverage of alloy, tin-free steel, and black plate exemplify the types of request that Commerce found easily justifiable.

Their inclusion in the August 25 proposal demonstrates the extent to which the US industry had failed to address negotiations previously and the degree to which their concerns were now being focused. Alloy steel is simply normal ("carbon") steel with small amounts of additional chemicals incorporated. It is more expensive than carbon steel, but less expensive than specialty steel. More important, it is usually made in the same plant, on the same equipment, as the corresponding carbon steel product. Domestic producers feared EC carbon steel producers would shift exports and production to the alloy products and destroy the growing high-profit alloy market in the United States. The 1981 EC share of the major alloy markets (structurals, hot- and cold-rolled sheet, and plate) was barely 1%, while those products accounted for over 3% of US consumption by tonnage (and more by value). For the same reason that US producers wanted to control EC entry to the alloy market, EC producers fought limitations, claiming there had been no injury yet in those products.

Tin-free steel and black plate also presented a strong threat of diversion if left uncovered. These high-value products are made in a tin mill along with tinplate, making the possibility of diversion real if tinplate was covered and the other products were not. The EC share of the US tin-free and black plate market was only 1.6% in 1981. Significantly, however, most EC exports of tin-free steel and black plate were from the Netherlands and West Germany, which had the least to fear from AD and CVD cases, and the least reason to restrain their exports.

The demand that consultations be held regularly with the objective of keeping EC import market share of products not subject to export licensing at or below the 1979-1981 level was particularly vexing. The US industry realized that it could not get total product coverage formally, and now sought to gain a commitment to keep the issue open for the life of the agreement. Important products to be dealt with in this manner would include the balance of specialty steel, hot-rolled alloy bars, and cold-finished bar, all of which the US government had tried and failed to get the EC to cover, for various reasons (mainly, too many mostly unsubsidized EC producers for easy consensus or control).

Roderick's proposal on the immunization issue was that the EC *not* be allowed to terminate any part or all of the system upon the filing of a single petition against any product; otherwise, the protection of each producer would be subject to the whims of any producer that might file a case. Instead, the EC would be entitled to terminate the arrangement only with regard to the product against which a petition had been filed, and only after an actual restraint was placed on imports of that product.

All of these demands were passed on to the EC, not as a proposal to reopen negotiations, but in recognition of the fact that the US industry was not going to accept the August 5 agreement without major changes. The EC continued to say publicly that it had a deal with the US government and that renegotiation was out of the question.

9 The End of the Game: September-October 21, 1982

From the time Roderick orally presented the US industry's demands on August 25 until the German government swallowed hard and accepted the arrangements on October 21, negotiations between, among, and within the various parties to the steel dispute were intense, continuous, exhausting, and often confusing. As an agreement became much more likely, the focus of discussion narrowed from the broad outlines to detailed specific language, a clause here or there that would protect one party or another from the potential perfidy of another.

Rather than present these negotiations in chronological order—which was confusing enough to participants, much less a reader—we will discuss the different issues in turn. We begin with the familiar issues of product coverage and export levels, a discussion that will also sketch the broad outlines of the negotiating process. We then review the boost given the arrangement by Congress, which passed a special law to provide enforcement authority, and touch briefly on several side-issues. Finally, we conclude with the two most difficult issues, which were not resolved until the last possible moment: immunization and pipe and tube.

PRODUCT COVERAGE AND LEVELS, AND CONSULTATIONS

The EC did not respond to the US industry's August 25 proposal for changes to the August 5 arrangement until late September, at which time the EC made enough concessions to keep the negotiations alive. In the interim the US and EC industries and the EC and member state governments had reviewed the final CVD determinations and their positions.

On September 20 Davignon sent his chef de cabinet Hugo Paemen to review the status of the negotiations with Under Secretary Olmer. Paemen sought Olmer's views on what modifications were necessary to gain the US industry's support. Olmer had anticipated that the EC would not stand on its position that an agreement already existed and would ask him to convince the industry to end the investigations. He had drawn up a written list of priorities from the requests provided by Roderick. Granting these priorities, Olmer told Paemen, would go a long way toward pacifying the petitioning steelmakers.

Commerce's priority list included only the additions to product coverage (alloy, tin-free steel and black plate, and sheet piling) and the reductions in market share in rail, bar, plate, and structurals. The latter two products were somewhat of a puzzle (as Table 8-1 shows); the reductions sought were only 6 and 4%. However, representatives of the industry had emphasized the importance of those reductions (perhaps, some Commerce officials guessed, for symbolic reasons), so the request was passed on as being relatively easily met.

Paemen accepted the list of priorities to take back to Brussels for study, and then proceeded to raise a technical product coverage issue. Detailed examination of the August 5 arrangement by EC steelmakers revealed that the US Customs definition of "plate" category (which was being used in the arrangement) included some of the product known commercially as "slab." The US industry had already acceded to the exclusion from the arrangement of most slab when they agreed to leave semifinished products outside its purview. Paemen explained that exclusion of coverage of this slab classified as plate was critical to one or more EC producers, who expected to supply Kaiser Steel's California plant with large quantities of slab—a plan that would be jeopardized if the arrangement was not amended. Olmer did not respond immediately, but before long a footnote excluding slab from the plate category was inserted into the arrangement.

On September 23, just two days after delivery of the priority list, Paemen phoned Olmer with major EC changes in position. Prior to this communication, many Commerce officials had a feeling of going through the motions solely because an effort was needed for political reasons. Now it seemed that full agreement was just around the corner. The EC would add alloy products to all the categories the US government had specified on the priority list, it would add terne plate to coated sheet, and it would create a separate category for sheet piling—all at 1981 levels, as requested. In addition, the market share of hot-rolled carbon bar would be reduced from the August 5 arrangement's 3.01% to 2.6% (the 1977-1982 average). The only items on the priority list not met were the addition of black plate and tin-free steel and the reduction of the rail market share, all said to be impossible because of the prominence of German and Dutch producers in those markets. No requests not on Commerce's priority list were met (requests that included the addition of alloy wire rod coverage, consultations on steel products not subject to licensing, and the immunization proposal).

Baldrige quickly conveyed the EC concessions to Roderick, who continued to speak for the US industry. Expecting to be greeted enthusiastically, Baldrige was surprised to hear Roderick reject the proposal out of hand and respond with yet more demands for alterations, demands in direct contradiction to the proposal Roderick had presented only a month earlier. Roderick now demanded, inter alia, that carbon and alloy products must be in separate categories, and that the hot-rolled sheet level must be reduced (even though the EC's latest proposal was *below* the August 25 request).

What had happened? Roderick implied that Olmer and Horlick had misunderstood his earlier proposal, that the levels he had then proposed pertained to carbon products only, and that he intended all along that alloy be included separately at 1981 levels. That explanation seemed unlikely, as DOC officials had confirmed the key points of Roderick's August 25 plan with the US Steel people working most closely with him. More likely was the explanation proffered informally by a US Steel official: Roderick's earlier plan had not been fully agreed to by the rest of the industry, and the other firms were now insisting on specific points of interest. In so doing, the several steel companies were taking advantage of the bargaining power their status as separate petitioners gave them. (In 1980 US Steel had been the sole petitioner, and the other companies had no direct control over the 1980 settlement; in 1982 some of the largest domestic steel producers filed, either individually or jointly, in large part to have a seat at the table in any settlement discussions—although Armco, a nonpetitioner, also participated in the 1982 negotiations.)

Hoping to avoid the confusion attendant upon his earlier proposal, on September 29 Roderick presented a detailed response to the EC's counterproposal in writing—the first written document concerning settlement passed from the steel industry to the government. The most important demand was that alloy be in separate categories, on the argument that without separate categories, the EC could upgrade its exports from carbon steel to alloy steel. That argument was valid, but there were no likely winning AD or CVD cases on alloy products. With carbon and alloy products under joint ceilings, the EC had agreed to pay for any increased alloy exports with less carbon exports; the US industry wanted to preclude increased alloy exports altogether. The struggle here was how far the industry would push, and how much the EC would, or could, give.

In addition to the alloy demand, the US industry made three other demands, labeled "key items," which were in direct contradiction to its earlier position. Consultations on imports of hot-rolled alloy bar now were not sufficient; that product must be subject to licensing at the 1981 level. This demand was difficult for the EC because the ITC had found no reasonable indication of injury to US producers from imports of hot-rolled alloy bar and had terminated all cases at a preliminary phase. In addition, the largest EC exporter of hot-rolled alloy bar to the United States exported almost nothing else and was not, it appeared, government owned nor controlled; and the remaining exports were split among over a dozen firms from six member states, most of whom were similarly specialized, independent firms.

The other two new demands concerned carbon hot-rolled sheet and carbon plate: both levels had to be reduced, as shown in Table 9-1.

TABLE 9-1. **US Industry September 29 Proposed Restraint Levels Compared to August 5 Arrangement Levels, Selected Products, in Percent of US Market**

	August 5 Arrangement	September 29 Proposal	Percent Reduction
Hot-rolled Carbon sheet	7.43	6.82	8.2
Carbon Plate	5.98	5.00	16.4

The magnitude of the required reductions (8.2% for hot-rolled sheet, 16.4% for plate) and the dispersion of exports of these products among EC producers made it difficult for the EC to seriously consider the proposal. The EC still maintained publicly that the August 5 arrangement was not open to renegotiation. The only product for which the level could be reduced from the August 5 level was hot-rolled carbon bar, where the predominance of heavily subsidized (and thus threatened by countervailing duties) British Steel made movement possible.

Roderick's written response also reiterated the importance of covering alloy wire rod (which had been omitted by the EC in part because it had not been on Commerce's priority list—but see below), tin-free steel, and black plate. The proposal had some positive notes: the industry now accepted the levels for rail (where it had previously demanded a reduction from the August 5 level, which the EC claimed was impossible because of the Germans and for hot-rolled carbon bar (reduced by the EC to 2.6% of the US market from 3.01% on August 6, versus the 2% demand by the US industry on August 25); and also accepted the August 5 level for structurals (10.9% instead of the requested 9.5%).

After analyzing Roderick's proposal for several days, during which time Commerce officials passed the details to the EC with little comment, Baldrige phoned Davignon both to express his anger at the US industry for backtracking and to ask Davignon to make some more concessions. Baldrige sought to be in a position to go back to Roderick and say, in effect, "You made these unreasonable demands in contradiction to your previous position. Despite the fact that it was next to impossible for the EC to come up with anything else, it made these additional concessions. Now you should agree to the arrangement."

Davignon did not disappoint. He said he could not agree to split carbon and alloy products into separate categories, but that perhaps some language on the necessity of avoiding diversion into alloy could be worked out. Davignon also agreed to bring hot-rolled alloy bar under export licensing. The other demands remained impossible for Davignon to meet. Baldrige relayed Davignon's concessions to Roderick. While Roderick gave no immediate indication as to whether the industry would accept an agreement with the product coverage and levels now on the table (October 1), neither did he cut off bargaining. With only three weeks remaining until the CVD results became final (and revocable only with great difficulty, if at all), the bargaining on product coverage and levels took a new direction. The US industry seemed to accept that it could not get all its desired results and followed Davignon's lead in trying to strengthen the language on consultations.

The August 5 arrangement provided for consultations quarterly and

at any other time at the request of either the ECSC or the US, to discuss any matters, including trends in the importation of steel products, which impair or threaten to impair the attainment of the objectives of this Arrangement.

By October 13 agreement was near on additional draft language, prepared by Commerce and EC officials:

In particular, consultations will be held in the event that imports from the ECSC of alloy products and of products not covered by the Arrangement show an increase indicating the possibility of diversion of trade from carbon to alloy products or from categories subject to export licensing to those not so subject. Should these consultations demonstrate that there has indeed been a diversion of trade which is such as to impair the attainment of the objectives of the Arrangement, then within 60 days of the request for consultations both sides will take the necessary measures for the products concerned in order to prevent such a diversion, including the creation of separate categories under the Arrangement for alloy products at the 1981 US market share level.

The convoluted language resulted from the differing aims of the negotiating partners. The EC—which was now unable or unwilling to limit exports of the products proposed to be subject to consultations and was not certain it could exercise greater control in the future—wanted to craft language that would make it seem to the US industry that the problem of unlicensed products had been solved, without commiting itself to take unsustainable actions in the future. The EC was not, however, averse to the US government taking unilateral steps to limit imports if need be. The Commerce Department was more concerned about the possibility of misleading the US industry with vague language and sought to get the EC to commit itself to take action if diversion occurred. An example of the negotiations leading to the above language was that the EC rejected Commerce's proposal that black plate and tin-free steel be specifically identified in the clause.

On October 14, one day before the ITC's vote on the final injury determination in the CVD cases and eight days before the absolute deadline for agreement, US Steel presented to Commerce officials the industry's position on the status of negotiations. US Steel, on behalf of the industry, accepted the scope and level of all products covered by the export-licensing requirement (except for wire rod, the producers of which were holding out for price protection as well as quotas). Specialty steel had by this time been deleted from the arrangement's licensing requirements because no meeting of the minds was possible. However, the consultation language was too weak for the industry's taste; it seemed to imply that separate licensing categories would not be created for uncovered products even if diversion were proved, and it set no benchmark by which to measure diversion. The alternative draft presented by the industry would have committed both the

EC and the United States to maintain the 1981 EC market share for *all* steel products, with imposition of export licenses by the EC one means of enforcement:

> Such consultations shall have the objective of maintaining the alloy products *and all other products not covered under the Arrangement* at an import level not to exceed the 1981 import penetration in order to assure that diversion . . . does not occur. Should these consultations demonstrate that there has been a diversion of trade . . . then within 60 days both sides . . . will take measures necessary to prevent such diversion, including the creation of separate categories under the Arrangement for alloy products and uncovered products at the 1981 market share level. [emphasis added]

This formulation was of concern to Commerce officials for two reasons: the EC would not accept it (meaning no agreement) and it ran very close to violating the guidelines given by Commerce lawyers to avoid possible antitrust liability—any export restraint must be mandatory in its terms so that the "foreign sovereign compulsion" doctrine would be an available defense. The prospect of actually setting, or being seen to set, EC steel export levels in informal private consultations was not appetizing.

October 15 brought the ITC's final CVD injury decisions. All remaining cases but two German cases were found to be causing injury, about as expected given the dismal state of the industry, the high level of subsidies, and the surge in imports. Chairman Eckes emphasized the commission's view (the law was not clear on this point) that the vote did not end the investigations; they did not end until October 21, when the ITC officially transmitted its decision to Commerce. An alternative view would have meant that the steel negotiations were dead; instead, a week remained.

On October 16 US and EC negotiators, all of them confused as to which products were subject to consultations and impatient with some of the convoluted and imprecise language of the arrangement, began to define classes of steel products precisely. The arrangement would deal with certain steel products (CSP) as defined in an appendix. Within CSP were two classes: arrangement products (AP), which were subject to export licensing, and CSP other than AP, which were subject to consultations. The definition clarified that specialty steel was totally excluded from the scope of the arrangement (it was not included in CSP). Equally important was the inclusion of semifinished steel (including slab) in the definition of CSP.

Thus, US Steel's contemplated importation of 3-million-plus tons of slab from British Steel for use in the Fairless mill became subject to the constraint on CSP other than AP negotiated at US Steel's insistence.

Also on October 16, the EC rejected the US industry's language implying that uncovered products could become subject to export licensing and that the objective of the consultations was to hold all uncovered products to 1981 levels or below. Two other changes to the language of the consultations clause were requested: that imports of consultation products would have to show a *significant* increase before consultations would be held, and that consultations would be held if it seemed that imports from third countries were replacing those limited by the arrangement. In some negotiations, it pays better to give up; Olmer finally agreed that the third country concern could be recognized to the extent of talking about it so long as there was no explicit or implicit commitment for the government to do anything about such potential displacement.

With only five days left, time pressure was on all sides. Commerce and US Steel officials developed a new idea by which the industry could be satisfied on several outstanding issues: a letter from Secretary Baldrige to the CEOs of the US steel companies that would expand upon the government view of the meaning of several clauses and the approach the Commerce Department would take in administering the arrangement and in consulting with the EC. This letter would not require an EC signature, only tacit EC approval. An early draft had Baldrige assuring the CEOs that

> if imports from the EC of CSP other than AP show a significant increase indicating the possibility of diversion of trade from AP to CSP other than AP, consultations will be held between the USG and the EC *with the objective of preventing diversion from the EC 1981 share of US apparent consumption.* [emphasis added]

The EC did not like the idea of putting anything in the CEO letter that would make it appear to its constituents that Davignon had made a private agreement with Baldrige. It therefore originally preferred that any reference to an objective of consultations be in the basic arrangement. It did not, however, object strenuously to Commerce's new formulation of the 1981 objective, which was much more ambiguous than the US steel industry's draft. The addition of slightly more ambiguity made the clause fully acceptable to the EC: "consultations will be held between the US and the EC with the objective of preventing diversion, *taking into account the ECSC 1981*

US market share levels" (emphasis added). Ultimately, this consultation language was accepted by the US steel industry, resulting in a much-watered-down commitment on CSP other than AP. The final agreement of the US industry to the consultation clause depended on the CEO letter, which laid out the approach the Commerce Department would take in consultations:

> With respect to Article 10, the US government will request consultation anytime that imports from Europe of CSP other than AP exceed the 1981 market share of US apparent consumption and any of your companies ask us to request such consultation.

The last piece of consultations clause that fell into place—and one of the last issues to be resolved by US, EC, and US industry negotiators prior to the draft arrangement being sent back to EC member states for approval—was the question of what the EC would do if diversion, however vaguely now defined, was "demonstrated" by consultations to have occurred. For alloy products, the question was resolved early on: the EC agreed to create separate licensing categories later in order to convince the US industry to drop its demand for separate categories now. However, the EC strenuously resisted parallel treatment of CSP other than AP. There were strong internal EC reasons for these products not being AP, reasons that would be just as strong in 1984 or 1985 as they were in 1982. Ambiguity was again called to the rescue, and the EC simply acknowledged the *possibility* of export licensing for CSP other than AP. Thus, the final shape of Article 10:

> Quarterly consultations shall take place between the ECSC and the US on any matter arising out of the operation of the arrangement. Consultations shall be held at any other time at the request of either the ECSC or the US to discuss any matters including trends in the importation of certain steel products which impair or threaten to impair the attainment of the objectives of this Arrangement.
> In particular, if imports from the ECSC of certain steel products other than Arrangement products or of alloy Arrangement products show a significant increase indicating the possibility of diversion of trade from Arrangement products or from carbon to alloy within the same Arrangement product, consultations will be held between the US and the ECSC with the objective of preventing such diversion, taking account of the ECSC 1981 US market share levels.

Should these consultations demonstrate that there has indeed been a diversion of trade which is such as to impair the attainment of the objectives of the Arrangement, then within 60 days of the request for consultations both sides will take the necessary measures for the products concerned in order to prevent such a diversion. For alloy Arrangement products, such measures will include the creation of separate products for purposes of Articles 3 and 4 at the 1981 US market share levels. For certain steel products other than Arrangement products, such measures may include the creation of products for purposes of Articles 3 and 4.

Consultations will also be held if there are indications that imports from third countries are replacing imports from the ECSC.

What does Article 10 mean in practice? First, we must note that, with respect to most products,* the diversion argument is a red herring; the EC steelmakers of CSP other than AP generally do not also manufacture APs. Rather, Article 10 uses diversion as a plausible pretext for a rather mealy-mouthed contingent export limitation. Some well-organized US producer of CSP other than AP (for example, cold-finished bar makers) mounted a campaign to get the Commerce Department to take some action under Article 10 almost before its ink was dry. Indeed, when it became public that US Steel was negotiating with British Steel to purchase slab for its Fairless mill, the Steelworkers Union wanted the Commerce Department to use Article 10 to prevent consummation of the deal. In sum, Article 10 traded immediate agreement and withdrawal of petitions for continuing disputes over steel trade—disputes in which there are no standards but political advantage, where the argument of diversion need not be examined critically, and in which the standards of fair trade may be abandoned for more promising returns from politicking.

Article 10 was not acceded to happily by many US government officials. It conflicted with their personal and professional preference for seeing standards of fairness combined with free competition, rather than political advantage, ruling international trade. Political necessity won out, however. The triumph of political expedience will be seen even more clearly later in the pipe and tube agreement.

---

*Exceptions include, for example, black plate. See page 77.

ENFORCEMENT

By late September all efforts of Commerce, Customs, and various interested party lawyers to find a palatable means for the United States to deny entry to unlicensed ECSC steel exports, which under the arrangement would be required to have a license, had failed. Without a legal means of denying entry, neither side would agree to the arrangement. The commitment made by Secretary of the Treasury Donald Regan to deny entry, on which the August 5 arrangement had been based, looked increasingly hollow without a legal justification. The only remaining alternative was new legislation. This was drafted on the sly by Commerce for Senator Heinz of Pennsylvania. After consultation with other executive branch agencies, Commerce agreed to draw the language of the legislation very narrowly so as not to provoke undue opposition from "free-traders," nor to provide executive authority that other industries might use to attempt to gain import protection.

SEC. 153. Title IV of the Tariff Act of 1930 (19 U.S.C. 1401 et seq.) is amended by adding after section 625 the following new section:

SEC. 626. (a) In order to monitor and enforce export measures required by a foreign government or customs union, pursuant to an international arrangement with the United States, the Secretary of the Treasury may, upon receipt of a request by the President of the United States and by a foreign government or customs union, require the presentation of a valid export license or other documents issued by such foreign government or customs union as a condition for entry into the United States of steel mill products specified in the request. The Secretary may provide by regulation for the terms and conditions under which such merchandise attempted to be entered without an accompanying valid export license or other documents may be denied entry into the United States.

(b) This section applies only to requests received by the Secretary of the Treasury prior to January 1, 1983, and for the duration of the arrangements.

The time limitation precluded use of the licensing requirement for any other steel deals (which would be practically impossible to negotiate before the deadline).

Congressional allies tied the legislation (drawn as an amendment to the Tariff Act of 1930) to the Treasury's appropriation bill and hoped to sneak it through without hearings or floor debate. They were successful in the Senate, and only slightly less so in the House; the amendment passed easily but only after biting criticism from Representative Frenzel of Minnesota. Frenzel objected to the ad hoc, special-situation change in trade law; the special treatment for the steel industry and the precedent set; and the cost of administering the VRA. He particularly objected to the procedure used to pass the bill through Congress. Nevertheless, he saw that opposition was futile and reluctantly accepted the amendment:

> Since we are determined to embark on this unwise policy, I hope that there will be an agreement, and that our steel industry will begin to work itself out of its problems. I am disappointed that we will adopt this bad policy, but we will make the best of it. I am far more ashamed of the procedure that produced it. (*Congressional Record*, H 8388, Oct. 1, 1982)

The major legal hurdle had now been cleared. If agreement could be reached by the negotiators, no legal concern stood in the way of a steel trade arrangement.

ARTICLE 8: THE SHORTAGE CLAUSE

In meetings immediately preceding the US steel industry's rejection of the August 5 arrangement, representatives of several firms objected to the inclusion of "price increases" as an indicator of shortage of steel products. The then-current draft of the shortage clause, derived from the August 5 document, read:

> If the US . . . determines that . . . the American steel industry will be unable to meet demand . . . (including substantial objective evidence such as allocation, extended delivery periods, significant price increase or other relevant factors) an additional tonnage shall be allowed.

Those representatives objected to the potential use of this authority by a future administration to limit steel price increases: the threat of opening up the import gates could be used to jawbone down steel prices. At a minimum, the US firms wanted the specific reference to prices deleted, as an indication that such use was not contemplated (even though that simple language change altered nothing substantively).

The EC claimed the shortage provision was of critical importance to several of its producers and was reluctant to delete the reference to price increases. Prices were perhaps the single best objective measure of shortages, one to which EC producers had immediate access; they resisted giving up that future bargaining strength. Again, resolution came in the form of a side-document recording an interpretation of the language in the arrangement itself. "Price increase" was deleted, and the below memorandum for the file from F. Lynn Holec, director, agreements compliance division, was prepared and became a part of the arrangement package:

> On October 14, 1982, the Commission of European Communities acceded to the US request to delete the reference to "significant price increase" in Article 8 of the EC–US Steel Arrangement. The US made this request because representatives of the US steel industry strongly objected to that reference in the August 5 Arrangement, fearing that a future US administration would use the threat of opening the gates to European imports as a means of limiting steel price increases. The DOC and EC agree that changes in price levels are an appropriate factor to consider in determining whether a shortage of a particular product exists. Where appropriate, we will consider changes in price levels as an "other relevant factor" under Article 8 of the Arrangement.

While the potential difficulties in interpreting and administering Article 8 were obvious to all, the prospect of a steel shortage seemed sufficiently remote in 1982 that potential problems were shrugged off by: Let's just hope the market recovers to the point where we can relish arguments over Article 8." A potential dilemma for the Commerce Department arose in early 1983 when it seemed possible that there would be an extended steel strike, which would force Commerce to choose between welcoming imports of EC steel—and infuriating the union—or keeping it out to the detriment of the wider economy. Timing was also critical, because of the two- to three-month lag between order and delivery. In order to ameliorate the effects of a strike, export limits would have to be lifted either before the strike occurred or immediately after it was declared. Then, because Article 8 provides for a six-month period of increased exports, if the strike ended quickly US steelmakers and workers would suffer increased imports nonetheless. Fortunately, the strike threat passed and Commerce was not confronted with a decision.

TRANSFER

While the US industry did not like the provisions in the August 5 arrangement allowing advance use, carryover, and transfer, the magnitude of flexibility allowed was so small that no serious objections were lodged—with one exception. Article 7 of the arrangement allowed the EC to transfer export tonnage from one category to another so long as the change in no category was greater than 5%. That 5% limitation was subject to increase upon US and EC agreement.

On October 14 the US industry's representative requested that the flexibility to increase the 5% limit be deleted. Because of perceived past abuse by various administrations (price controls, both formal and jawboning, and failure to control imports satisfactorily), many member firms distrusted the government almost as much as they distrusted the EC. They saw the possibility of a future administration allowing massive shifts of imports in *their* product line to punish them for one reason or another (such as for price increases).

At this late stage of the negotiations the EC was not willing to make what seemed to it insignificant changes in the text of the arrangement. It noted that the arrangement already gave control over the increase of the 5% limit to the US government; how that control would be exercised was up to the government (although it demanded that some real flexibility be retained). The solution was obvious—a written statement of intent by Baldrige to the CEOs in the CEO letter:

> With respect to Article 7(a) of the Arrangement on certain steel products, the US Government will not agree to increase the percentage limit beyond 10 percent.

TRANSITION PERIOD

The need for some understanding on the transition period (the span from the date of agreement on the arrangement to the date of imposition of export controls) arose because of US industry concern that EC producers would take advantage of the time required to implement the licensing system to export all available steel. The August 5 arrangement had addressed the problem in a general manner, the idea being that any abnormal surge in exports during the transition period would be subtracted from permissible exports after licensing was imposed:

> The ECSC shall ensure that in regard to exports effected between 1st August and 30th September 1982, aberrations from seasonal trade patterns will be accommodated in the ensuing licensing period.

The US industry and Commerce staff were not satisfied with the vagueness of the formulation. Soon after the rejection of the August 5 arrangement, Commerce drafted tighter language that would specify that any excess of exports in August and September above arrangement market share levels would be deducted thereafter. The EC rejected this language as overly restrictive, and proposed in early October interpretative language as opaque as any seen in these negotiations:

> The elements to be taken into account to see whether Article 2(3) is to be applied are:
>
> (1)   Has there been a "surge" in relation to the annual curve of the seasonal distribution of exports over the three previous years,
>
> (2)   Has such surge been preceded or not by abnormally low exports given the seasonal trade curve during the previous months of the year?

The EC was trying to establish the average export tonnage for the August-September period over 1979-1981 as the baseline by which to measure 1982 exports over the period, with consideration given to whether January-July 1982 exports were abnormally low. This baseline would be quite high in comparison to arrangement levels, as it was set in relation to 1979-1981 levels (higher than 1981, in turn higher than the arrangement) and on tonnage, not market share (1979-1981 tonnages would result in very high market shares because of the depressed 1982 market). These higher transition period levels were not acceptable to the US industry. All did agree that the arrangement's export licensing should begin November 1.

After learning of the unacceptability of its proposed benchmark, the EC tried another formulation. It argued that August and September 1982 should be dropped out of the arrangement altogether, and that the transition period should apply only to October (since October was then well under way). United States negotiators refused to drop two months, and reminded the EC of Paemen's earlier agreement that the EC would "payback" for excesses of early 1982.

Since agreement on a baseline for August-September 1982 proved impossible, a new tactic was undertaken. The EC's concern was that if exports in the transition period exceeded the baseline, it would have to distribute the quota reductions among EC steel producers. The EC was already having great difficulty securing agreement on export quotas; agreement on reductions would be complicated and difficult. It therefore polled its steel firms to estimate the level of exports in the transition period and aimed the benchmark at that specific tonnage figure.

Once again, rather than change the language from the August 5 arrangement, the negotiators relied upon a side-understanding. The United States and EC would agree on what precise tonnage level would constitute "an aberration from seasonal trade patterns." The negotiations became rather simple: the EC sought a figure that would leave it room for error in its estimate of actual imports, and the United States sought a figure low enough to be justifiable to its steel producers (who would be presented with a tonnage figure and be left to determine its relationship to historic levels themselves—under time pressure and with more important issues pending).

The EC's initial offer was a benchmark of 1 million metric tons. Because this number seemed to be out of historical proportion, Commerce negotiators gradually chipped away until the EC agreed to 968,000 metric tons (1,069,000 net tons). Based on estimates of US apparent consumption, even that tonnage level would result in EC arrangement market shares above historical levels, as Table 9-2 demonstrates. United States industry representatives did notice that the tonnage benchmark was a bit high and protested. However, only a few days of negotiation remained, and they chose to accept the transition level and concentrate on the more important remaining issues: immunization and pipe and tube.

IMMUNIZATION

The August 5 arrangement would have deterred the filing of import relief petitions by allowing the EC to terminate any or all of the arrangement if a petition was filed against any covered product. In addition, Article 2(c)

TABLE 9-2. **Estimated Levels of EC Exports of Arrangement Products to the United States, US Apparent Consumption of Those Products, and EC Market Share, August-October various years (net tons and percent)**

Period	Estimated EC Exports of Arrangement Products	Estimated US Apparent Consumption	Estimated EC Market Share
1979-81	1,301,000	17,170,000	7.6
1982	1,069,000	12,000,000	8.9

was interpreted by the EC as giving it the same termination authority (and therefore deterrent power) in the event a petition was filed against *non-covered* products. (The US industry saw Article 2(c) as requiring the US government to consult with the EC before taking actions against non-covered products, and felt no constraints on its freedom to file cases against noncovered products. The divergence of views on 2(c) was never explored in negotiations subsequent to August 5.) The relevant language was:

> 2(b)   If . . . investigations under [list of U.S. laws] are initiated or petitions filed or litigation instituted with respect to Arrangement products, the ECSC shall be entitled to terminate the Arrangement with respect to some or all of the Arrangement products after consultations with the US.
>
> 2(c)   If . . . any of the above mentioned proceedings or litigation is instituted in the USA against steel products imported from the Community which are not subject to the Arrangement and which substantially threaten its objective, the ECSC and the US, before taking any other measure, shall consult to consider appropriate remedial measures.

Neither the US steel industry nor the EC was satisfied with this formulation. The industry's concerns were stronger, but it did not have, at first, a reasonable alternative formulation. In both its August 25 oral and its follow-up September 29 written proposal, it sought to make the EC's right to terminate the system contingent on additional US import restrictions resulting from the filing of a petition, *not* the filing of the petition itself. To the EC, this formulation was grossly inadequate; it foresaw the industry filing petitions to harass the EC into making various kinds of concessions during the life of the arrangement—for example, to get the arrangement extended beyond 1985. The US industry also sought to limit the EC's termination power to affected products only.

The first change in the August 5 language came at EC behest on October 11 (Article 2 negotiations were concentrated into a very short period near the October 21 deadline). The EC had heard that Bethlehem Steel had an antitrust case ready to file against EC steelmakers, and therefore demanded explicit insertion of antitrust cases in the list of actions, petitions, and litigation warranting termination. This demand was rather easily met (by inserting "including antitrust" after "litigation"), as was a similar request from US Steel. At the time, US Steel was planning to file a petition under Section 301 of the Trade Act of 1974, claiming that an agreement between Japan and the EC caused damage to the US industry and requesting quotas on Japanese steel imports as relief. It wanted assurance that even though the petition would claim EC misdeeds, its filing would not affect the arrangement. The arrangement was already clear on this point, and the CEO letter once again proved its usefulness. Baldrige reassured the skittish CEOs that

> If a petition is filed on an Arrangement product, Article 2(b) would apply. Because Arrangement products are defined in Article 1 as exports to or destined for consumption in the US of products described in Article 3(a) originating in the Community, Article 2(b) does not apply to petitions or litigation that will not affect US imports of those products from the EC.

On October 14 the US steel industry, in its report to Commerce officials on outstanding points of contention, presented a new idea to deal with the "Korf scenario." (Willy Korf, who owned wire rod manufacturers in both the United States and the EC, had been independent of the major US and EC producers. The "Korf scenario" referred to the potential for someone beyond the influence of the major producers to file petitions and kill the system.) Under the plan, the EC would be entitled to terminate the arrangement with respect to only those products affected by the litigation and only if consultations determined that "such petition or litigation is likely to have an adverse effect on the ability of the parties to continue the Arrangement with respect to the products affected by such petition or litigation." It was not specified who had to make the determination.

The EC, as expected, was not receptive to any proposed limitation on its right to terminate the system. Peace in the valley demanded, from their perspective, the deterrent effect of total destruction; that threat would keep intra-US industry pressures strong and prevent a pick-off of one product at a time. Commerce therefore floated a new idea: that the EC retain the right of total destruction, but that the required consultations after a petition was filed but before the EC could terminate would

> take into account the nature of the petitions or litigation, the identity of the petitioner, the amount of trade involved, the scope of relief sought, the degree to which the petition or litigation threatens the objectives of the Arrangement, and other relevant factors.

It was hoped that specification of these factors would reassure the US industry that the EC would not take advantage of the first opportunity presented to escape its commitments.

The US industry accepted the new idea of linking the EC's right of termination to the identity of the petitioner, but did not accept a mere specification of petitioner's identity as a consideration to be taken into account. It was giving up some winning AD and CVD cases on the verge of completion and wanted concrete assurance that US firms beyond its control would not destroy the system for which it had paid so much. Working together with Commerce officials, US industry negotiators put together another proposal on the afternoon of October 18, which was immediately cabled off to Brussels.

The new proposal would allow the EC to terminate the arrangement: with respect to the product affected by any litigation or petition, regardless of the identity of the petitioner; with respect to any or all products, if one of the current petitioners (those agreeing to the arrangement by withdrawing pending petitions and/or litigation) filed a petition or litigation on any arrangement product; and with respect to any or all products, "fifteen days after consultations resulting in a determination that a petition or litigation concerning an Arrangement product is likely to have an adverse affect on the ability of the parties to continue the Arrangement." Again, it was not specified *who* was to make the determination. The 15-day waiting period had been added to allow the major US producers time to "discuss" the potential consequences of a petition or litigation with smaller members of the industry.

Late October 18, the answer came back from Brussels: the EC could not accept the constraints on its power to terminate the system. Brussels saw the precondition of consultations "resulting in a determination" as requiring a joint US-EC decision, and would not accept US government control over its deterrent. Despite US negotiators' disavowal of the necessity for a joint decision—Olmer explained that the words were purposely ambiguous—the EC required either the clear elimination of the joint decision feature or more ambiguity. Since the steel industry would not agree to any substantial increase in the EC deterrent, the answer was found in a slight increase in ambiguity. As time ran out, the three parties agreed to change "resulting in a determination" to "it is determined." The change to the passive voice without identification of who would be making the determination sealed agreement: the EC believed it retained unilateral authority to terminate the arrangement in whole or part upon filing by a Korf of a petition dealing with only one AP, while the US steel industry was able to claim that that authority is not unilateral. The text as finally agreed upon reads as follows:

> b) If during the period in which the Arrangement is in effect, any such investigations* or investigations under Section 201 of the Trade Act of 1974, Section 232 of the Trade Expansion Act of 1962, or Section 301 of the Trade Act of 1974 (other than Section 301 petitions relating to third country sales by US exporters) are initiated or petitions filed or litigation (including antitrust litigation) instituted with respect to the Arrangement products, and the petitioner or litigant is one of those referred to in article 2a) [those withdrawing petitions], the ECSC shall be entitled to terminate the Arrangement with respect to some or all of the Arrangement products after consultations with the US, at the earliest 15 days after such consultations.
>
> If such petitions are filed or litigation commenced by petitioners or litigants other than those referred to in the previous paragraph, or investigations initiated, on any of the Arrangement products, the ECSC shall be entitled to terminate the Arrangement with respect to the Arrangement product which is the subject of the petition, litigation or investigation after consultations with the US, at the earliest 15 days after such consultations. In addition, if during the consultations it is determined that the petition, litigation or investigation threatens to impair the attainment

---

*With respect to any Section 337 investigation, the parties shall consult to determine the basis for the investigation. [This footnote was added to permit Section 337 investigations based on alleged patent or trademark infringement, but not AD and CVD or antitrust cases disguised as Section 337 actions.]

of the objectives of the Arrangement, then the ECSC shall be entitled to terminate the Arrangement with respect to some or all Arrangement products, at the earliest 15 days after such consultations.

These consultations will take into account the nature of the petitions or litigation, the identity of the petitioner or litigant, the amount of trade involved, the scope of relief sought, and other relevant factors.

c) If, during the term of this Arrangement, any of the above mentioned proceedings or litigation is instituted in the USA against certain steel products as defined in Article 3b) imported from the Community which are not Arrangement products and which substantially threaten its objective, then the ECSC and the US, before taking any other measure, shall consult to consider the appropriate remedial measures.

One result of the final week's negotiations on Article 2(b) was to give sharper definition to Article 2(c). It would be illogical for the EC to now maintain that it has total freedom, after consultations, to terminate any or all of the arrangement if a case was filed on a CSP other than AP, given the limitation on that power contained in 2(b) with respect to the more important arrangement products. However, the EC never acknowledged a changed interpretation of 2(c); one can only hope that paragraph is never tested.

PIPE AND TUBE

The August 5 arrangement resolved nothing in pipe and tube except that the Commerce Department and the EC committed themselves to reach some kind of agreement by September 15. This deadline allowed plenty of time for withdrawal of the pipe and tube petitions before Commerce determinations on the cases hardened positions; the preliminary CVD ruling on small diameter welded pipe from West Germany and France was due on October 4 (covering only 10,608 net tons of the 567,015 net tons of steel pipe and tube imported from the EC in the first quarter of 1982).

Early on in Commerce–EC discussions on pipe and tube, EC negotiators proposed that the US and EC pipe and tube producers get together directly and work out an agreement. This suggestion probably arose from the unwillingness of the EC pipe and tube producers to have the commission do their negotiating for them. (As noted, the pipe and tube industry was not subject to the ECSC and the major producers had been profitable and without need of government assistance.) Commerce officials were not averse to the prospect of direct industry-to-industry talks in principle, but were wary of antitrust consequences. Therefore, any talks were put off until the US Department of Justice could render an opinion on their legality.

On July 30, 1982, William Baxter, assistant attorney-general for the antitrust division of the Justice Department, delivered his views on the risk of antitrust liability inherent in the steel negotiations and on whether representatives of the two industries could participate in the talks. On the negotiations themselves, Baxter noted that so long as the negotiations resulted in mandatory export controls, the foreign sovereign compulsion doctrine would preclude liability for the conduct compelled by those controls. The risk posed by participation of private parties was that a court might conclude that the resulting restraints were private in nature, and that the respective governments had merely rubber-stamped that private agreement. Baxter therefore counseled strongly against direct industry-to-industry negotiations.

Baxter's strongest warning concerned the possibility that negotiations would not result in mandatory EC export controls:

> I should add that even the scenario in which the United States and European producers meet only with their respective governments is not entirely without risk. If such talks should result in no agreement at all or in a voluntary rather than EC-imposed mandatory curtailment of exports by the European producers, it might be argued that any reduction in exports that thereafter occurred was the result of an agreement between the European and the US producers, with government negotiators acting (whether or not intentionally) as a "conduit" for communications between the two groups of producers. The possibility of no agreement being reached cannot, in any way I see, be protected against; but our officials should take care neither to encourage nor accept any arrangement involving nonmandatory restraints by the European producers.

Baxter's view set the basic outlines of the Commerce Department's position: any pipe and tube agreement should involve mandatory EC export restraint in one form or another. It also determined the negotiating forum: no direct industry-to-industry talks. Thus, the EC's demand for those talks had to be met with indirection.

The US industry had said very little about pipe and tube other than that it must be dealt with and that the appropriate market share level was 5.9%. Therefore, going into the first round of negotiations on September 8 the US position was simply a demand for mandatory restraint at 5.9%. The structure of the September 8 meeting was, however, not that simple.

The EC claimed it was unable to convince its producers to accept restraint; those producers wanted to hear the argument for restraint from the US government directly. Baxter's rules made this difficult, but not impossible: senior executives of EC pipe manufacturers accompanied government officials as industry advisors (a not uncommon phenomenon in international negotiations) and Commerce officials spoke only to the EC member state government officials (the EC itself was present at the initial meeting only as an observer).

All input to the meeting by the EC "advisors" was through whispers to EC member state government officials, who passed on the comments or questions. The Commerce Department had accepted as an industry advisor an ex-official of US Steel Corporation; his presence at the meeting served to represent the US pipe and tube industry. While the requisite format made the formal session somewhat ludicrous, it did not diminish its importance.

The challenge the meeting presented to Commerce officials was great: there were no compelling reasons why EC pipe and tube producers should accept restraint since there was little chance of successful AD and CVD cases. The US industry (producers of both basic carbon steel products and pipe and tube) was demanding that the interests of independent EC pipe and tube makers be sacrificed as the price for its assent to an agreement ameliorating the dilemma of EC carbon steel producers. It seemed to DOC officials that the task of convincing, or forcing, the EC pipe industry to go along could be accomplished only in Europe. Private warnings had been passed to Commerce officials that the EC pipe and tube producers were very touchy; any tactless statements about the political reality might cause them to walk out.

Under Secretary Olmer therefore undertook a limited task: to be sure the EC producers understood the full extent of the disaster in the US pipe and tube market, and to advance possible reasons for them to accept restraint. Olmer made two arguments. The first was meant to appeal to baser instincts: the market was in disarray, prices had collapsed, and if steps were not taken to restore order *all* pipe and tube producers, US *and* European, would suffer low prices. The second was an appeal to public spirit and long-range self-interest: unless the EC steel producers accepted restraint, chances were that the US Congress would unilaterally impose steel quotas (including pipe and tube quotas). Olmer recited a familiar litany of growing protectionist pressure exacerbated by the recession: the auto domestic content bill, steel quota bills pending, the powerful and protectionist Congressional Steel Caucus. Olmer did not present the fundamental US goal—that the EC must impose mandatory export controls to limit EC pipe and tube to 5.9%—to the EC steel executives. Rather, after the formal session he met privately with EC Commission and member state representatives and told them his goal. They did not comment on Olmer's proposal but took the matter back to Brussels for consideration.

The initial reaction of the EC pipe and tube makers to the general thrust of Olmer's remarks was predictable. They were impressed by his presentation, but argued that the proper way to restore order in the market was to allow market forces to work, as they already were. Their US orders had evaporated, they said, and exports from Europe were declining rapidly. As for heading off protectionism, they agreed on the danger of protectionism and challenged the Reagan administration to fight it rather than concede defeat. The EC steelmakers also reiterated the arguments that pipe and tube is different than basic carbon steel and should not be dealt with in the same fashion as there had been no unfair trade.

Within a week of the September 8 meeting, secondhand reports to Commerce officials indicated that the 5.9% figure was within a point or so of being acceptable to the EC Commission. However, it was reported that the German government was strongly opposed to the concept of mandatory export controls.

On September 20 Olmer and Paemen discussed pipe and tube in the context of their review of the US industry's August 25 proposal. Olmer repeated the necessity for mandatory export controls on pipe and tube; any other kind of agreement would potentially put EC pipe and tube producers (and possibly other participants in the negotiations) afoul of US antitrust laws. Paemen seemed less concerned about a legal need for mandatory export controls and suggested an alternative plan. The plan essentially called for a statistical monitoring system with an antisurge mechanism. The EC would announce a target level, monitor subsequent exports, and if a tendency to exceed the announced limits was observed, take action to prevent it. The suggestion took Olmer and his aides by surprise; it seemed precisely the kind of system against which Baxter and Commerce staff had cautioned. Rather than totally rejecting the suggestion and reiterating the need for compulsion, however, the Commerce officials told Paemen that the US Justice Department probably would not approve such a program.

On October 4 Commerce issued its preliminary determinations that neither the French nor the German steel pipe and tube manufacturers had received significant subsidies (in Commerce terminology, the subsidies were de minimis, less than 0.5% of the sales value). Backs stiffened in Europe.

In the first week of October, the EC presented its first specific, written proposal on pipe and tube. Because the Germans would not allow any mandatory export restraint and the US Justice Department would not allow any voluntary export restraint, the only solution was no export restraint at all. The EC stated it was confident that, simply because of market conditions, there was no reason for concern about EC exports of pipe and tubing to the United States in the foreseeable future. The EC proposed to demonstrate the truth of that statement to the satisfaction of the US industry by immediate communication to the US government of EC pipe and tubemakers' detailed information on existing orders and projected exports, followed by ex-factory shipments on a monthly basis until 1985. The proposal went on to contemplate that

> At the end of a six month period, consultations will take place to review the situation and to draw any appropriate conclusions. Consultations can be requested at any time by either of the two parties in the event of any particular problem arising. Before these consultations with the US take place, the Commission will examine, together with the industry and representatives of the Member States, the development of exports and will, if the need arises, make appropriate proposals.

The US industry immediately rejected the proposal, which contained no hint of export restraint, much less confirming the 5.9% level and the restrictions on product mix now demanded. After that rejection was fully understood by the EC, with time running out, the EC made another proposal privately on October 12, and publicly on October 15.

The second EC proposal began with a point of agreement: that the United States and EC agree that diversion from arrangement products to pipe and tube should be avoided. Diversion thus became the pretext for an arrangement on pipe and tube, providing a (very) thin veneer of respectability. The proposal went on to state that the EC believed that so long as export of pipe and tube does not exceed the 1979-1981 average (5.9%), no diversion will have taken place. After provision for the previously proposed information exchange came the guts of the proposal: What would happen if diversion *did* occur? Either party could request consultation. Then,

> If estimates show that the 79/81 average might be exceeded, consultations will take place in order to find an appropriate solution.

If after 60 days no solution is found, the party who asked for the solution will be free:

—either to take with respect to the exports and within the existing legislative and regulatory framework measures which it considers necessary, in accordance with international obligations

—or put an end to its obligations which it has assumed under the steel Arrangement.

The proposal was surprising to the United States, to say the least. The EC was formally proposing that if there was diversion to pipe and tube, the United States could either take any actions it liked within existing rules (which it was already free to do, and required by law to do), or it could terminate the arrangement on CSP. Astonishingly, the proposal contemplated that if EC saw a problem in pipe and tube trade, asked for consultations, and failed to reach a solution satisfactory to it, the EC could terminate the basic arrangement. Termination of the basic arrangement by any party before its expiration date was the last thing the US industry wanted; it was fighting to limit the EC's termination rights under the immunization clause and wanted to extend the expiration date to 1987 or beyond. Commerce officials had a difficult time understanding the thinking behind this plan. EC representatives explained that the threat of termination of the basic carbon steel arrangement would give the EC the leverage over its independent pipe and tube makers to comply with informal export restraints, but neither the US industry nor the Commerce Department would buy that logic.

Hoping to build a silver lining into the latest EC cloud, the Commerce Department sought to build on the framework presented by the EC by beefing up the "what-if" clause and, pursuant to US industry demands, adding a product mix requirement. On October 14 (with one week left before the deadline), Commerce cabled Brussels a new proposal based on the EC's October 12 draft. The key charges were the definition of diversion:

The EC is of the opinion that such diversion will not take place insofar as exports of pipe and tube to the US do not exceed the 1979-81 average share of annual US apparent consumption *in any major pipe and tube category*,

and the what-if clause:

Within 60 days of the request for the consultations, the EC *will take appropriate governmental measures* to prevent or end any identified diversion. In aid of these measures, the US may take any necessary measures within the legislative or regulatory framework with respect to pipe and tube from the EC. [emphasis added]

The EC was being asked to *commit* to act ("EC *will*"). The latter phrase on enforcement was added at US industry insistence that the United States enforce the pipe and tube arrangement as well as the basic arrangement. A US proposal to change existing drafts of two other critical documents not yet discussed here was made simultaneously: the letters to the Secretary of the Treasury requesting that the new law on prohibition of entry (Section 626) be invoked.* Previous drafts of these letters referred to "an arrangement" and left sufficient room for an interpretation that the request covered CSP other than AP in the event that the EC subsequently imposed export licensing on any of those products. The new draft referred to "arrangements," contemplating that Section 626 would be invoked if necessary with respect to pipe and tube.

The US industry was the first to object to the October 14 Commerce Department formulation as altogether too vague. What, it asked, might an "appropriate governmental measure" be? With only five days remaining, it reiterated that without some certainty on pipe and tube there could be no deal. Commerce therefore returned to the EC and sought to strengthen the pipe and tube arrangement further still. The what-if clause was tightened to assure US enforcement. Instead of the EC taking "appropriate measures" with US help allowed, diversion would call for enforced export licensing under new Section 626. Upon a finding of diversion,

> both parties will take, with respect to pipes and tubes, measures within their legislative and regulatory frameworks necessary and adequate to invoke Section 626.

While this Commerce draft was nearly acceptable to the US steel industry (it still wanted tighter language on product mix, including identification of specific product categories), the EC continued to reject any formulation that could commit it to take any specific actions to limit pipe and tube exports in the future. Its big concession on October 17 was to agree that, if consultations did not resolve any problems, both sides *will* take not-yet-agreed-upon action. As to that action, the reference to Section 626 was said to be an infringement on EC sovereignty. The Germans strongly rejected a definition of diversion as exceeding the 1979-1981 EC market share *of any*

*See page 89 infra.

major product category. They said that to have a chance of not exceeding 1979-1981 averages by product category an internal EC export-allocation system would be needed, which was politically impossible, and that sublimits would inevitably mean less total exports. Commerce suggested that perhaps sublimits were not needed; the market conditions calling for consultations could be expanded to include "a distortion of product mix," a much less severe approach than including specific product group limitations in the definition of diversion.

October 18 was planned to be the day of completion of negotiations so that the EC would have time to finish its internal approval process prior to the final October 21 deadline. That day saw many drafts shuffled back and forth between the negotiators in Washington and Brussels. Working together in the afternoon of October 18, Commerce and US industry negotiators drafted a proposal on all outstanding issues for the EC and cabled it to Brussels. The proposals would require EC movement if they were to be accepted. Because of the time needed for the EC to complete its internal approval process prior to the required October 21 petition withdrawal, the EC wanted to distribute a completed arrangement (to which the commission had acceded) by 4:00 P.M., October 19, Brussels time (10 A.M. Washington time). Upon distribution of the arrangement to the member states, a written approval process would begin. If any member state rejected the proposal, the EC Council of Ministers, the highest authority, would meet Wednesday morning, October 21, pushing the time limit (5:15 P.M. October 21 Washington time) to the limit. On cabling of the proposal, US negotiators were not optimistic that a settlement would be reached. The EC still insisted that if consultations on pipe and tube were unsuccessful, it could terminate *both* steel arrangements, and that it could not commit to impose mandatory pipe and tube export controls. The final resolution of the status of CSP other than AP and the immunization issue also remained.

In these last few days before the October 21 deadline, negotiators informally discussed a novel and creative idea that, if accepted, could allow a settlement rather easily. It was argued that if the various documents were crafted carefully enough, the United States could *unilaterally* (that is, without EC export licensing) enforce the 5.9% market share mentioned in EC's proposed monitoring/surge mechanism. The EC had agreed to make the Section 626 request letter apply to *both* arrangements, and it was now suggested that EC export licensing was not required for US enforcement. Section 626 merely required a request (which was in hand), an international agreement (also in hand), and "export measures required by a foreign government or customs union." It was suggested that the EC's statistical monitoring system would constitute those "export measures," and that if needed the United States could "require the presentation of a valid export license or other documents issued by such foreign government or customs union as a condition for entry into the United States of steel mill products specified in the request" even though *no such documents existed.*

The purpose of such a sneaky means of potential US government enforcement was that the EC had to avoid painting itself into a corner with unsustainable promises of future mandatory pipe and tube export controls. Various negotiators saw that the US industry was most interested in US government enforcement and hoped that the creation of a future enforcement mechanism would convince the US industry to sign off without requiring any substantive EC commitment. In pursuit of this goal, other means of US enforcement were also floated, such as Section 301 of the Trade Act of 1974.

The Commerce response was that it was up to the EC to control pipe and tube trade. United States negotiators were anxious to maintain a facade of free trade. They were not eager to take action in direct contradiction to the GATT nor give encouragement to other domestic industries seeking import protection. Because it was the EC seeking escape from the application of US trade laws, US officials felt that the onus for action should fall on that side of the Atlantic.

The afternoon of October 18, following receipt of the US proposals in Brussels, brought what US negotiators thought to be the climactic confrontation: Olmer and Baldrige (with Roderick standing by) spoke to Davignon in Brussels, and believed that all outstanding issues had been satisfactorily resolved (which at this time included, in addition to pipe and tube, the precise immunization formula). All that remained was for the specific language to be cabled from Brussels. Near midnight, the EC's Washington representatives brought a letter cabled from Brussels that, instead of sealing the deal as expected, put the whole affair in doubt. Any agreement reached that afternoon had been illusory. The EC would not give up the right to terminate "the arrangement" (that is, kill the basic arrangement if consultations under the pipe and tube arrangement failed to resolve a complaint), nor would it recognize a pipe and tube product category limit. It did agree that if diversion occurred, "either party will take . . . measures which it considers necessary. In doing so, both will act in a complementary fashion to prevent diversion," an opaque reference to the possibility of invoking Section 626 if needed.

On October 19 Commerce negotiators (less this author, who was at home with his wife having a baby) reached Davignon in Brussels. Davignon communicated that the EC would accept a vague product category limit in the pipe and tube arrangement, and would drop its insistence that it be permitted to terminate the basic arrangement if the pipe and tube consultations should fail. These concessions were reportedly made following intense internal EC discussions, primarily concerning burden-sharing.

The final critical language was, at last, agreed upon by Baldrige, Davignon, and Roderick:

> The Communities are of the opinion that such a diversion will not take place in so far as annual exports of pipes and tubes to the US do not exceed the 1979-81 average share of annual US apparent consumption. . . .
>
> If estimates based on the above information and projections of US apparent consumption of pipes and tubes show that the 1979-81 average described in paragraph A might be exceeded *or that a distortion of the pattern of US–EC trade* is occurring within the pipe and tube sector, consultations between the EC and the US will take place in order to find an appropriate solution. If after 60 days no solution has been found each party *will take*, within its legislative and regulatory framework, *measures which it considers necessary*. In doing so *both parties will act in a complementary fashion in order to prevent diversion.* (emphasis added)
>
> If during the period in which this Arrangement is in effect, any petitions seeking import relief under US law, including countervailing duty, antidumping duty, Section 337 of the Tariff Act of 1930, Section 201 of the Trade Act of 1974, Section 301 of the Trade Act of 1974, or Section 232 of the Trade Expansion Act of 1962, are filed or investigations initiated or litigation (including antitrust litigation) instituted with respect to pipe and tube products, and the petitioner or litigant is one of those referred to in paragraph A above or in Article 2(a) of the Arrangement concerning certain steel products, the ECSC shall be entitled to terminate this Arrangement after consultation with the US, at the earliest 15 days after such consultations.
>
> If such petitions are filed or litigation commenced by petitioners or litigants other than those referred to in the previous paragraph, or investigations initiated, on pipe and tube products, the ECSC will be entitled to terminate this Arrangement if during consultations with the US it is determined that the petition, litiga-

tion or investigation threatens to impair the attainment of the objectives of this Arrangement. These consultations will take into account the nature of the petitions or litigation, the identity of the petitioner or litigant, the amount of trade involved, the scope of the relief sought, and other relevant factors.

A close reading of the pipe and tube arrangement and an understanding of its background clearly reveal that the EC committed to almost nothing. It merely agreed to enter consultations and to take measures *it considered necessary*. Later US assertions that the EC had in fact committed to maintain a 5.9% market share (and actions, including an embargo on imports of pipe and tube from the EC, based on that assertion) ignored both the letter of the arrangement and its long and arduous negotiating history.

THE EC APPROVES THE ARRANGEMENT

While the EC member states did ultimately approve the arrangements, it was not without some change to the structure. Much of the dispute over ultimate EC approval concerned internal EC allocation of export quotas. The Germans in particular demanded preference in the allocation of export licenses and ultimately, it appears, received it.

More importantly, the Germans demanded that the title of the previously labeled "Arrangement on EC Export of Pipes and Tubes to the USA" be changed—that the arrangement be conveyed titleless as a letter from Davignon to Baldrige. The EC tried to resist but could not, given German resistance and time pressure. The EC asked the United States to make the change, but US negotiators saw the change as weakening the EC commitment, delinking the two arrangements, and disassociating pipe and tube from Section 626 (both the statute and the request letters to the Secretary of the Treasury were keyed to "arrangements"). Notwithstanding this refusal, the EC, without informing the United States, altered the language of the pipe and tube commitment in the papers actually sent to the United States—*after* both sides had announced agreement. (During the latter stages of the negotiations, the various documents had resided in Commerce's word-processing system, and agreement was based on that Commerce text.) The US steel firms had no opportunity to see the change until after they had withdrawn their petitions.

What does this change mean? First of all, the switch led to some meaningless language—the EC document refers to the possible "termination of the present exchange of letters" if a petition is filed (instead of "terminate the arrangement"). Second, because the United States had used the term "arrangement" in the confirmation letter to Davignon, the two sides had not agreed on precisely the same language. This bothered State Department lawyers, who requested that Commerce renegotiate the language after the fact to obtain conformity (an impossibility). Third, the applicability of Section 626 was made slightly more arguable (although by oversight the EC's letter to Secretary Regan requesting the invocation of that law still refers to the plural "arrangements"). Most serious, it indicates that the Germans really intended the language to mean what it says: that the EC will consult and do what *it* considers necessary (which could be nothing).

10   Summary and Conclusion

Underneath the tremendous complexity of the negotiations leading up to the 1982 steel trade arrangements lies a simple truth: the US steel industry was able to use leverage created by a judicialized system of trade dispute settlement to force its government to accept a political system of import protection that it would otherwise have refused. The Reagan administration in 1982 would not have acceded to steel import quotas of any sort merely because of domestic industry pressure, but when EC pressure for protection from US laws was added, it could not refuse.

At the same time that the legalistic import protection procedures (for which the US steel industry had vigorously lobbied) enabled the industry to achieve a quota arrangement with the EC, those same procedures prevented the industry from achieving its real goal (control on imports from all countries) and limited its ability to control imports from EC producers. Faced with judicial review of their determinations, Commerce and the ITC had little alternative to finding no subsidization or injury in many of the cases against the EC. Similarly, successful cases against the two leading non-European exporters of steel to the United States (Japan and Canada) were impossible and low dumping and subsidy rates were found for South Korea. By defining what is unfair, US trade laws also defined what *is* fair, making steps to limit those fair imports in some sense illegitimate. Global quotas could not be forced by the US industry on the administration via the unfair trade laws, which in fact worked against that goal by clearing major sources of steel imports of wrongdoing.

The interaction of the US trade laws and the negotiating process was particularly ironic for Commerce Department officials who were both conducting the investigations and handling the negotiations. Clearly, the negotiations would have been a lot easier if the Germans had been found to be subsidized, because it would have given the German producers an incentive to agree to the quotas. But the very laws sought by the US industry precluded any manipulation of the cases even if the Commerce officials had been so minded. The mechanics of the entire negotiation reflected this law-driven quality.* The negotiations reached a first crisis point shortly before the preliminary countervailing duty determination, then subsided until another crisis point prior to the deadline for suspension agreements under the statute. This led to a tentative agreement just prior to the release of the antidumping results, and to the withdrawal of the cases literally minutes before DOC, under the statute, would have been required to impose the duties.

The critical importance of the legal structure should not allow us to overlook the effects of both domestic and international politics. The US government is fundamentally a political body; officials are loyal to the president and their party as well as to a broader national interest. Where necessary, those officials will accept or even advocate policies that they and their staff might not believe optimal for the problem faced, but which will pacify other political actors (such as members of Congress, unions, and blocs of voters) whose support is necessary for other issues. The sacrifice of the desired steel trade policy (enforcement of fair trade, allowing the market to determine the fate of the US industry) to broader national security, economic, and electoral goals is an example of the political manner in which the US government deals with trade/industrial policy that advocates of a more consciously interventionist industrial policy should closely study.

---

*See John H. Jackson, "Crumbling Institutions of the Liberal Trade System," *Journal of World Trade Law* 12, No. 2 (March-April 1978), 98–101.

APPENDIXES

A   August 5 Arrangement*

Draft ECSC Letter to US

Dear,

As we have discussed, the Commission of the European Communities, on behalf of the European Coal and Steel Community, is prepared to restrain exports of certain steel products to the United States.

It is our understanding that, in conjunction with such action by the ECSC, the United States Government is prepared to undertake certain other actions vis-a-vis trade in these products.

The elements of our program and a description of the complementary US actions are set forth in the attached (the "Arrangement").

In entering into this Arrangement, the ECSC does not admit to having bestowed subsidies on the manufacture, production or exportation of the products the subject of the countervailing duty petitions to be withdrawn or that any such subsidies have caused any material injury in the USA.

This Arrangement is entered into without prejudice to the rights of the US Government and of the European Community under the GATT. We understand that the US Government recognizes the [foreign policy and trade implications] of this Arrangement vis-a-vis trade in covered products with the European Community, and will be fully cognizant of this fact in exercising its discretionary authority under par. 337, 201, 103 and par. 232 with regard to such products and shall do so only after consultations with the Community.

The independent forecaster for the purposes of Article 5 of the Arrangement shall be Data Resources, Inc.

With regard to Article 2(a) (1) the ECSC accepts that the condition shall be regarded as fulfilled concerning the Section 301 petitions listed in Appen-

---

*The original August 5 arrangement contained various technical appendixes that have been omitted for the purpose of this book.

dix A which cover both Arrangement products and other products, if the U.S. takes measures to cease its investigations relating to, and takes no action upon, the Arrangement Products.

[Consultations between the ECSC and the US will be held in 1985 to review the desirability of extending and possibly modifying the Arrangement.]

[Given the magnitude of trade in the pipe and tube sector (not included in the Arrangement] both parties recognise that it is necessary to ensure that trade distortions in that sector do not arise which would undermine this Arrangement and create problems in that sector. To this end, discussions should therefore rapidly be engaged on this matter and it is on this assumption that the Department of Commerce agrees to this Arrangement. The nature and the objective of these discussions are not the same as for the Arrangement, nor will be the legal means by which the results will be attained. Both parties undertake to use their best endeavours to resolve this matter by 15 September.]

I look forward to hearing from you at your early convenience.

Yours faithfully,

Attachment

ARRANGEMENT

concerning trade in certain steel products between the European Coal and Steel Community (hereinafter called "the ECSC") and the United States (hereinafter called "the US").

1. Basis of the Arrangement

Recognizing the policy of the ECSC of restructuring its steel industry including the progressive elimination of state aids pursuant to the ECSC State Aids Code; recognizing also the process of modernization and structural change in the United States of America (hereinafter called the "USA"); recognizing the importance as concluded by the OECD Steel Committee in 197[8] of restoring the competitiveness of OECD steel industries; and recognizing, therefore, the importance of stability in trade in certain steel products between the European Community (hereinafter called "the Community") and the USA;

The objective of this Arrangement is to give time to permit restructuring and therefore to create a period of trade stability. To this effect the ECSC* shall restrain exports to or destined for consumption in the USA of products described in Article 3 originating in the Community (such exports hereinafter called "the Arrangement products") for the period 1st October 1982 to 31st December 1985.

The ECSC shall ensure that in regard to exports effected between 1st August and 30th September 1982, aberrations from seasonal trade patterns will be accommodated in the ensuing licensing period.

2. Condition—Withdrawal of petitions; new petitions

a) The entry into effect of this Arrangement is conditional upon:

   (1) the withdrawal of the petitions and termination of all investigations concerning all countervailing duty, antidumping duty and Section 301 petitions listed in Appendix A at the latest fifteen days before the commencement of the Initial Period; and

   (2) receipt by the US at the same time of an undertaking from all such

---

*To the extent that the Arrangement products are subject to the Treaty establishing the European Economic Community, the terms "ECSC" should be substituted by "the EEC."

petitioners not to file any petitions seeking import relief under US law, including countervailing duty, antidumping duty, Section 301 of the Trade Act of 1974 (other than Section 301 petitions relating to third country sales by US exporters) or Section 337 of the Tariff Act of 1930, on the Arrangement products during the period in which this Arrangement is in effect.

b) If during the period in which this Arrangement is in effect any such investigations* or investigations under Section 201 of the Trade Act of 1974, Section 232 of the Trade Expansion Act of 1962, or Section 301 of the Trade Act of 1974 relating to third country sales by US exporters are initiated or petitions filed or litigation instituted with respect to the Arrangement products, the ECSC shall be entitled to terminate the Arrangement with respect to some or all of the Arrangement products after consultations with the US.

c) If, during the term of this Arrangement, any of the above mentioned proceedings or litigation is instituted in the USA against steel products imported from the Community which are not subject to this Arrangement and which substantially threaten its objective, then the ECSC and the US, before taking any other measure, shall consult to consider appropriate remedial measures.

3. Product description

The products covered by the present Arrangement are the following steel products:

Hot-rolled sheet and strip
Cold-rolled sheet
Plate
Structurals
Wire rods
Hot-rolled bars
Coated sheet
Tin plate
Rails
Stainless steel sheet and strip and stainless steel plate

---

*With respect to any Section 337 investigation, the parties shall consult to determine the basis for the investigation.

as described and classified in Appendix B by reference to corresponding Tariff Schedules of the United States Annotated (TSUSA) item numbers and EC NIMEXE classification numbers.

4. Export limits

a) For the period 1st October 1982 to 31 December 1983 (hereinafter called "the Initial Period") and thereafter for each of the years 1984 and 1985 export licenses shall be required for the Arrangement products. Such licences shall be issued to Community exporters for each product in quantities no greater than the following percentages of the projected US Apparent Consumption (hereinafter called "export ceilings") for the relevant period:

Product	Percentage
Hot-rolled sheet and strip	7.43
Cold-rolled sheet	5.15
Plate	5.98
Structurals	10.90
Wire rods	4.29
Hot-rolled bars	3.01
Coated sheet	3.32
Tin plate	2.20
Rails	8.90
Stainless steel sheet and strip and stainless steel plate	4.08

For the purposes of this Arrangement, "US Apparent Consumption" shall mean shipments (deliveries) minus exports plus imports, as described in Appendix D.

b) Where products covered by this Arrangement exported from the Community to the USA are subsequently reexported therefrom, without having been subject to substantial transformation, the export ceiling for such products for the period corresponding to the time of such reexport shall be increased by the same amount.

c) For the purposes of this Arrangement the USA shall comprise both the US Customs Territory and US Foreign Trade Zones: in consequence the entry into the US Customs Territory of Arrangement products which have already entered into a Foreign Trade Zone shall not then be again taken into account as imports of Arrangement products.

5. Calculation and revision of US Apparent Consumption forecast and of export limits

The US, in agreement with ECSC, will select an independent forecaster which will provide the estimate of US Apparent Consumption for the purposes of this Arrangement.

For the Initial Period, a first projection of the US Apparent Consumption by product will be established as early as possible and in any event before 15th September 1982. A provisional export ceiling for each product will then be calculated for that period by multiplying the US Apparent Consumption of each product by the percentage indicated in Article 4 for that product. These figures for projected consumption will be revised in October 1982, February, May, August and October of 1983, by the said independent forecasters, and appropriate adjustments will be made to the export ceilings for each product taking into account licences already issued under Article 4.

The same procedure will be followed to calculate and revise the US Apparent Consumption and export ceilings for 1984 and for 1985, the first projection being established by the independent forecasters by 1st October of 1983 and 1984 respectively.

In February of each year as from 1984, adjustments to that year's export ceiling for each product will be made for differences between the forecasted US Apparent Consumption and actual US consumption of that product in the previous year or (in February 1984) in the Initial Period.

6. Export licenses

a) By Decisions to be published in the Official Journal of the European Communities the ECSC will require an export licence for all Arrangement products. Such export licenses will be issued in a manner that will avoid abnormal concentration in exports to the USA taking into account seasonal trade patterns. The ECSC shall take such action, including the imposition of penalties, as may be necessary to make effective the obligations resulting from the export licenses. The ECSC will inform the US of any violations concerning the export licences which come to its attention and the action taken with respect thereto.

Export licences will provide that shipment must be made within a period of three months.

Export licences will be issued against the export ceiling for the Initial Period or a specific calendar year as the case may be. Export licences may be used as early as 1st December of the previous year within a limit of eight (8) percent of the ceiling for the given year. Export licences may not be used after 31st December of the year for which they are issued except that licences not so used may be used during the first two months of the following year with a limit of eight (8) percent of a) the export ceiling of the previous year or b) eighty (80) percent of the export ceiling of the Initial Period, as the case may be.

b) The ECSC will require that the merchandise shall be accompanied by an authenticated copy of a part of the licence containing the details set out in Appendix C. The US shall require presentation of such copy as a condition for entry into the USA of the Arrangement products. The US shall prohibit entry of such products not accompanied by such copy of a valid licence.

7. Technical adjustments

a) The specific product export ceilings provided for in Article 4 may be adjusted by the ECSC. Adjustments to increase the volume of one product must be offset by an equivalent volume reduction for another product for the same period. Notwithstanding the preceding sentences, no adjustment may be made under this paragraph

—which results in an increase or a decrease in a specific product limitation under Article 4 by more than five (5) percent by volume for the relevant period, or

—between the first nine product categories listed in Article 3 (carbon steel products) and the last such product category (stainless steel products).

The ECSC and the US may agree to increase the above percentage limit.

b) Normally, only one change in a specific product export ceiling in a given year or the Initial Period may be made by an adjustment under the preceding paragraph or use of licences in December of January/ February under Article 6(a). Accordingly, changes in a given year or the Initial Period by use of more than one of those three provisions may be made only upon agreement between the ECSC and the US.

8. Short supply

On the occasion of each quarterly consultation provided for in Article 10 the US and the ECSC will examine the supply and demand situation in the USA for each of the products listed in Appendix B. If the US in consultation with the ECSC determines that because of abnormal supply or demand factors, the American steel industry will be unable to meet demand in the USA for a particular product (including substantial objective evidence such as allocation, extended delivery periods, [significant price increase] or other relevant factors) an additional tonnage shall be allowed for such product or products by a special issue of licences limited to 10 percent of the ECSC's unadjusted export ceiling for that product or products. In extraordinary circumstances as determined by the US in consultation with the ECSC the US will increase the allowable level of special licences.

Each authorized special issue export licence shall be so marked and must be used within 180 days after the start of the quarter when that special issue began.

9.   Monitoring

The ECSC will within one month of each quarter and for the first time by 31st January 1983 supply the US with such non-confidential information on all export licences issued for Arrangement products as is required for the proper functioning of this Arrangement.

The US will collect and transmit quarterly to the ECSC all non-confidential information relating to export licenses received during the preceding quarter in respect of the Arrangement products, and relating to actions taken in respect of Arrangement products for violations of customs laws.

10.   General

Quarterly consultations shall take place between the ECSC and the US on any matter arising out of the operation of the Arrangement. Consultations shall take place at any other time at the request of either the ECSC or the US, to discuss any matters, including trends in the importation of steel products, which impair or threaten to impair the attainment of the objectives of this Arrangement.

11. Scope of the Arrangement

This Arrangement shall apply to the US Customs Territory (except as otherwise provided in Article 4(c)) and to the territories to which the Treaty establishing the ECSC as presently constituted applies on the conditions laid down in that Treaty.

12. Notices

For all purposes hereunder the US and the ECSC shall be represented by and all communications and notices shall be given and addressed to:

for the ECSC
The Commission of the European Communities
(Directorates General for External Relation (I) and
Internal Market (III))
rue de la Loi, 200
1049 Brussels, BELGIUM
Tel: 235.11.11
Telex: 21877 COMEU B

for the US
US Department of Commerce
International Trade Administration
Washington, D.C. 20230
Tel: 202/377.17.80
Telex: 892536 USDOC WSH DAS/IA/ITA

APPENDIX A

List of countervailing duty (CVD), antidumping (AD) and Section 301 petitions to be withdrawn:

—The 19 CVD cases on certain steel products from the UK, France, Italy, the FRG, Belgium, Luxembourg, and the Netherlands initiated by DOC on February 1, 1982.

—The 17 AD cases on certain steel products from France, the UK, Italy, Belgium, the Netherlands, Luxembourg, and the FRG initiated by DOC on February 1, 1982.

—The CVD cases on wire rod from Belgium and France initiated by DOC on March 1, 1982.

—The AD cases on stainless steel sheet and strip from the FRG and France initiated by DOC on May 17, 1982 and June 1, 1982, respectively.

—The CVD case on rail from the EC filed with DOC by the petitioners on July 22, 1982.

—The AD cases on rail from France, the FRG, and the UK filed with DOC by the petitioners on July 22, 1982.

—The Section 301 investigations on stainless steel sheet and strip and plate from France, the UK and Belgium.

B   EC and US Letters of October 21

COMMISSION
OF THE
EUROPEAN COMMUNITIES

The Honorable Malcolm BALDRIGE
Secretary
Department of Commerce
Washington D.C. 20230
U.S.A.

Brussels, October 21, 1982

Dear Mr. Secretary,

As we have discussed, the European Coal and Steel Community and the European Economic Community (EC) are prepared to restrain certain steel exports to the United States.

It is our understanding that, in conjunction with such action by the EC, the United States Government is prepared to undertake certain other actions vis-a-vis trade in these products.

The elements of our program and a description of the complementary U.S. actions are set forth in the enclosed text (the "Arrangement").

In entering into this Arrangement, the EC does not admit to having bestowed subsidies on the manufacture, production or exportation of the products that are the subject of the countervailing duty petitions to be withdrawn or that any such subsidies have caused any material injury in the U.S.A. Neither does it admit that its enterprises have engaged in dumping practices which are the subject of the anti-dumping duty petitions to be withdrawn or that any such practices have caused any material injury in the U.S.A.

This Arrangement is entered into without prejudice to the rights of the U.S. Government and of the EC under the GATT.

- 2 -

We understand that the US Government recognizes the implications of this Arrangement vis-a-vis trade in certain steel products as defined in the Arrangement with the EC for international competitiveness, national economic and security interests, and trade in capital goods, and will be fully cognizant of these implications in exercising its discretionary authority under section 337 of the Tariff Act of 1930, section 201 and 301 of the Trade Act of 1974, section 232 of the Trade Expansion Act of 1962, and section 103 of the Revenue Act of 1971 with regard to such products and shall do so only after consultations with the EC.

The independent forecaster for the purposes of Article 5 of the Arrangement shall be Data Resources, Inc.

Consultations between the EC and the US will be held in 1985 to review the desirability of extending and possibly modifying the Arrangement.

I look forward to hearing from you at your early convenience.

Yours faithfully,

E. DAVIGNON
Vice President

THE SECRETARY OF COMMERCE
Washington, D.C. 20230

2 1 OCT 1982

Vicomte Etienne Davignon
Vice-President of the European Communities
Rue de la Loi 200
1049 Brussels
Belgium

Dear Mr. Vice-President:

I have received your letter of October 21, 1982, worded as follows:

> "The Honorable Malcolm Baldrige
> Secretary of Commerce
> Washington, DC 20230 USA
>
> Dear Mr. Secretary:
>
> As we have discussed, the European Coal and Steel Community
> and the European Economic Community (EC) are prepared to
> restrain certain steel exports to the United States.
>
> It is our understanding that, in conjunction with such action
> by the EC, the United States Government is prepared to
> undertake certain other actions vis-a-vis trade in these
> products.
>
> The elements of our program and a description of the
> complementary U.S. actions are set forth in the enclosed text
> (the "Arrangement").
>
> In entering into this Arrangement, the EC does not admit to
> having bestowed subsidies on the manufacture, production or
> exportation of the products that are the subject of the
> countervailing duty petitions to be withdrawn or that any such
> subsidies have caused any material injury in the U.S.A.
> Neither does it admit that its enterprises have engaged in
> dumping practices which are the subject of the antidumping
> duty petitions to be withdrawn or that any such practices have
> caused any material injury in the U.S.A.

This Arrangement is entered into without prejudice to the rights of the U.S. Government and of the EC under the GATT.

We understand that the U.S. Government recognizes the implications of this Arrangement vis-a-vis trade in certain steel products as defined in the Arrangement with the EC for international competitiveness, national economic and security interests, and trade in capital goods, and will be fully cognizant of these implications in exercising its discretionary authority under section 337 of the Tariff Act of 1930, sections 201 and 301 of the Trade Act of 1974, section 232 of the Trade Expansion Act of 1962, and section 103 of the Revenue Act of 1971 with regard to such products and shall do so only after consultations with the EC.

The independent forecaster for the purposes of Article 5 of the Arrangement shall be Data Resources, Inc.

Consultations between the EC and the U.S. will be held in 1985 to review the desirability of extending and possibly modifying the Arrangement.

I look forward to hearing from you at your early convenience.

Yours faithfully,

Enclosure"

I have the honor to confirm the agreement of the U.S. Government with the contents of your letter.

Very truly yours,

Malcolm Baldrige

Secretary of Commerce

C   October 21 Arrangement*

ARRANGEMENT

concerning trade in certain steel products between the European Coal and Steel Community (hereinafter called "the ECSC") and the United States (hereinafter called "the US").

1. Basis of the Arrangement

Recognizing the policy of the ECSC of restructuring its steel industry including the progressive elimination of state aids pursuant to the ECSC State Aids Code; recognizing also the process of modernization and structural change in the United States of America (hereinafter called the "USA"); recognizing the importance as concluded by the OECD of restoring the competitiveness of OECD steel industries; and recognizing, therefore, the importance of stability in trade in certain steel products between the European Community (hereinafter called "the Community") and the USA;

The objective of this Arrangement is to give time to permit restructuring and therefore to create a period of trade stability. To this effect the ECSC† shall restrain exports to or destined for consumption in the USA of products described in Article 3 a) originating in the Community (such exports hereinafter called "the Arrangement products") for the period 1st November 1982 to 31st December 1985.

The ECSC shall ensure that in regard to exports effected between 1st August and 31st October 1982, aberrations from seasonal trade patterns of Arrangement products will be accommodated in the ensuing licensing period.

2. Condition—Withdrawal of petitions; new petitions

a) The entry into effect of this Arrangement is conditional upon:

---

*The original October 21 Arrangement contained various technical appendixes that have been omitted for the purposes of this book.

†To the extent that the Arrangement products are subject to the Treaty establishing the European Economic Community (the EEC), the term "ECSC" should be substituted by "the EEC."

(1) the withdrawal of the petitions and termination of all investigations concerning all countervailing duty and antidumping duty petitions listed in Appendix A at the latest by 21st October 1982; and

(2) receipt by the US at the same time of an undertaking from all such petitioners not to file any petitions seeking import relief under US law, including countervailing duty, antidumping duty, section 301 of the Trade Act of 1974 (other than Section 301 petitions relating to third country sales by US exporters) or Section 337 of the Tariff Act of 1930, on the Arrangement products during the period in which this Arrangement is in effect.

b) If during the period in which this Arrangement is in effect any such investigations* or investigations under Section 201 of the Trade Act of 1974, Section 232 of the Trade Expansion Act of 1962, or Section 301 of the Trade Act of 1974 (other than Section 301 petitions relating to third country sales by US exporters) are initiated or petitions filed or litigation (including anti-trust litigation) instituted with respect to the Arrangement products, and the petitioner or litigant is one of those referred to in Article 2 a), the ECSC shall be entitled to terminate the Arrangement with respect to some or all of the Arrangement products after consultations with the US, at the earliest 15 days after such consultation.

If such petitions are filed or litigation commenced by petitioners or litigants other than those referred to in the previous paragraph, or investigations initiated, on any of the Arrangement products, the ECSC will be entitled to terminate the Arrangement with respect to the Arrangement product which is the subject of the petition, litigation or investigation, after consultation with the US at the earliest 15 days after such consultation. In addition, if during these consultations it is determined that the petition, litigation or investigation threatens to impair the attainment of the objectives of the Arrangement, then the ECSC shall be entitled to terminate the Arrangement with respect to some or all Arrangement products at the earliest 15 days after such consultations.

---

*With respect to any Section 337 investigation, the parties shall consult to determine the basis for the investigation.

These consultations will take into account the nature of the petitions or litigation, the identity of the petitioner or litigant, the amount of trade involved, the scope of relief sought and other relevant factors.

c) If, during the term of this Arrangement, any of the above mentioned proceedings or litigation is instituted in the USA against certain steel products as defined in Article 3 b) imported from the Community which are not subject to this Arrangement and which substantially threaten its objective, then the ECSC and the US, before taking any other measure, shall consult to consider appropriate remedial measures.

3. <u>Product description</u>

a) The products are:

Hot-rolled sheet and strip
Cold-rolled sheet
Plate
Structurals
Wire rods
Hot-rolled bars
Coated sheet
Tin plate
Rails
Sheet piling

as described and classified in Appendix B by reference to corresponding Tariff Schedules of the United States Annotated (TSUSA) item numbers and EC NIMEXE classification numbers.

b) For purposes of this Arrangement, the term "certain steel products" refers to the products described in Appendix E.

4. <u>Export Limits</u>

a) For the period 1st November 1982 to 31st December 1983 (hereinafter called "the Initial Period") and thereafter for each of the years 1984 and 1985 export licences shall be required for the Arrangement products.

Such licences shall be issued to Community exporters for each product in quantities no greater than the following percentages of the projected US Apparent Consumption (hereinafter called "export ceilings") for the relevant period:

Product	Percentage
Hot-rolled sheet and strip	6.81
Cold-rolled sheet	5.11
Plate	5.36
Structurals	9.91
Wire rods	4.29
Hot-rolled bars	2.38
Coated sheet	3.27
Tin plate	2.20
Rails	8.90
Sheet piling	21.85

For the purposes of this Arrangement, "US Apparent Consumption" shall mean shipments (deliveries) minus exports plus imports, as described in Appendix D.

b) Where Arrangement products imported into the USA are subsequently reexported therefrom, without having been subject to substantial transformation, the export ceiling for such products for the period corresponding to the time of such reexport shall be increased by the same amount.

c) For the purposes of this Arrangement the USA shall comprise both the US Customs Territory and US Foreign Trade Zones. In consequence the entry into the US Customs Territory of Arrangement products which have already entered into a Foreign Trade Zone shall not then be again taken into account as imports of Arrangement products.

5. Calculation and revision of US Apparent Consumption forecast and of export limits

The US, in agreement with ECSC, will select an independent forecaster which will provide the estimate of U.S. Apparent Consumption for the purposes of this Arrangement.

For the Initial Period, a first projection of the US Apparent Consumption by product will be established as early as possible and in any event before 20th October 1982. A provisional export ceiling for each product will then be calculated for that period by multiplying the US Apparent Consumption of each product by the percentage indicated in Article 4 for that product. These figures for projected consumption will be revised in December 1982, February, May, August and October of 1983, by the said independent forecasters, and appropriate adjustments will be made to the export ceilings for each product taking into account licences already issued under Article 4.

The same procedure will be followed to calculate and revise the US Apparent Consumption and export ceilings for 1984 and for 1985, the first projection being established by the independent forecasters by 1st October of 1983 and 1984 respectively.

In February of each year as from 1984, adjustments to that year's export ceiling for each product will be made for differences between the forecasted US Apparent Consumption and actual US apparent consumption of that product in the previous year or (in February 1984) in the Initial Period.

6. Export licenses
a) By Decisions and Regulations to be published in the Official Journal of the European Communities the ECSC will require an export licence for all Arrangement products. Such export licences will be issued in a manner that will avoid abnormal concentration in exports of Arrangement products to the USA taking into account seasonal trade patterns. The ECSC shall take such action, including the imposition of penalties, as may be necessary to make effective the obligations resulting from the export licences. The ECSC will inform the US of any violations concerning the export licences which come to its attention and the action taken with respect thereto.

Export licences will provide that shipment must be made within a period of three months.

Export licences will be issued against the export ceiling for the Initial Period or a specific calendar year as the case may be. Export licences may be used as early as 1st December of the previous year within a limit of eight (8) percent of the ceiling for the given year. Export licences may not be used after 31st December of the year for which they are issued except that licences not so used may be used during the first two months of the following year with a limit of eight (8) percent of the export ceiling of the previous year or of eight (8) percent of eighty-six (86) percent of the export ceiling of the Initial Period, as the case may be.

b) The ECSC will require that the Arrangement products shall be accompanied by a certificate, substantially in the form set out in Appendix C, endorsed in relation to such a licence. The US shall require presentation of such certificate as a condition for entry into the USA of the Arrangement products. The US shall prohibit entry of such products not accompanied by such a certificate.

7. Technical adjustments

a) The specific product export ceilings provided for in Article 4 may be adjusted by the ECSC with notice to the US Adjustments to increase the volume of one product must be offset by an equivalent volume reduction for another product for the same period. Notwithstanding the preceding sentences, no adjustment may be made under this paragraph which results in an increase or a decrease in a specific product limitation under Article 4 by more than five (5) percent by volume for the relevant period.

The ECSC and the US may agree to increase the above percentage limit.

b) Normally, only one change in a specific product export ceiling in a given year or the Initial Period may be made by an adjustment under the preceding paragraph or use of licences in December or January/February under Article 6(a). Accordingly, changes in a given year or the Initial Period by use of more than one of those three provisions may be made only upon agreement between the ECSC and the US.

8. Short supply

On the occasion of each quarterly consultation provided for in Article 10 the US and the ECSC will examine the supply and demand situation in the USA for each of the products listed in Appendix B. If the US in consultation with the ECSC determines that because of abnormal supply or demand factors, the US steel industry will be unable to meet demand in the USA for a particular product (including substantial objective evidence such as allocation, extended delivery periods or other relevant factors) an additional tonnage shall be allowed for such product or products by a special issue of licences limited to 10 percent of the ECSC's unadjusted export ceiling for that product or products. In extraordinary circumstances as determined by the US in consultation with the ECSC the US will increase the allowable level of special licences.

Each authorized special issue export licence and certificate derived therefrom shall be so marked. Each such licence must be used within 180 days after the start of the quarter when that special issue began.

9. Monitoring

The ECSC will within one month of each quarter and for the first time by 31st January 1983 supply the US with such non-confidential information on all export licences issued for Arrangement products as is required for the proper functioning of this Arrangement.

The US will collect and transmit quarterly to the ECSC all non-confidential information relating to certificates received during the preceding quarter in respect of the Arrangement products, and relating to actions taken in respect of Arrangement products for violations of customs laws.

10. General

Quarterly consultations shall take place between the ECSC and the US on any matter arising out of the operation of the Arrangement. Consultations shall be held at any other time at the request of either the ECSC or the US to discuss any matters including trends in the importation of certain steel products which impair or threaten to impair the attainment of the objectives of this Arrangement.

In particular, if imports from the ECSC of certain steel products other than Arrangement products or of alloy Arrangement products show a significant increase indicating the possibility of diversion of trade from Arrangement products to certain steel products other than Arrangement products or from carbon to alloy within the same Arrangement product, consultations will be held between the US and the ECSC with the objective of preventing such diversion, taking into account the ECSC 1981 US market share levels.

Should these consultations demonstrate that there has indeed been a diversion of trade which is such as to impair the attainment of the objectives of the Arrangement, then within 60 days of the request for consultations both sides will take the necessary measures for the products concerned in order to prevent such a diversion. For alloy Arrangement products, such measures will include the creation of separate products for purposes of Articles 3 and 4 at the ECSC 1981 US market share levels. For certain steel products other than Arrangement products, such measures may include the creation of products for purposes of Articles 3 and 4.

Consultations will also be held if there are indications that imports from third countries are replacing imports from the ECSC.

11.   Scope of the Arrangement

This Arrangement shall apply to the US Customs Territory (except as otherwise provided in Article 4(c)) and to the territories to which the Treaty establishing the ECSC as presently constituted applies on the conditions laid down in that Treaty.

12.   Notices

For all purposes hereunder the US and the ECSC shall be represented by and all communications and notices shall be given and addressed to:

> for the ECSC
> The Commission of the European Communities
> (Directorates General for External Relations (I) and for
> Internal Market and Industrial Affairs (III))
> rue de la Loi, 200
> 1049 Brussels, BELGIUM
> Tel: 235.11.11
> Telex: 21877 COMEU B
>
> for the U.S.
> U.S. Department of Commerce
> Deputy Assistant Secretary for Import Administration

International Trade Administration
Washington, D.C. 20230
Tel: 202/377-17-80
Telex: 892536 USDOC WSH DAS/IA/ITA

APPENDIX A

List of countervailing duty (CVD) and antidumping duty (AD) petitions* to be withdrawn:

—CVD petitions, filed on January 11, 1982, by (1) United States Steel Corporation, (2) Bethlehem Steel Corporation, and (3) Republic Steel Corporation, Inland Steel Company, Jones and Laughlin Steel, Inc., National Steel Corporation, and Cyclops Corporation concerning certain steel products from Belgium, France, the Federal Republic of Germany, Italy, Luxembourg, the Netherlands, the United Kingdom, and the European Communities.

—AD petitions, filed on January 11, 1982, by (1) United States Steel Corporation, and (2) Bethlehem Steel Corporation concerning certain steel products from Belgium, France, the Federal Republic of Germany, Italy, Luxembourg, the Netherlands, and the United Kingdom.

—CVD petitions, filed on February 8, 1982, by Atlantic Steel Corporation, Georgetown Steel Corporation, Georgetown Texas Steel Corporation, Keystone Consolidated, Inc., Korf Industries, Inc., Penn Dixie Steel Corporation and Raritan River Steel Company concerning carbon steel wire rod from Belgium and France.

—CVD petitions, filed on May 7, 1982, by United States Steel Corporation concerning carbon steel welded pipe from France, the Federal Republic of Germany and Italy.

—CVD petition, filed on September 3, 1982, by CF and I Steel Corporation concerning steel rails from the European Communities.

—AD petitions, filed on September 3, 1982, by CF and I Steel Corporation concerning steel rails from France, the Federal Republic of Germany and the United Kingdom.

---

*For purposes of this Arrangement, the term "petitions" covers all matters included in the petitions filed on the dates listed, whether or not the DOC initiated investigations on the products or countries concerned.

D Pipe and Tube Arrangement

COMMISSION
OF THE
EUROPEAN COMMUNITIES

Brussels, 21 October 1982

The Honorable Malcolm BALDRIGE
Secretary
Department of Commerce
Washington, D.C. 20230
U.S.A.

Dear Mr. Secretary,

In our conversations we agreed to have an exchange of letters concerning pipes and tubes establishing the following.

A. It has been agreed during negotiations on trade in steel mill products between the European Communities (EC) and the United States (US) that for the duration of the Arrangement negotiated for those products diversions of trade from steel products described in Appendix B of the steel Arrangement towards pipes and tubes should be avoided. The US Government wishes trade in the tube sector to be examined at this stage. The Communities are of the opinion that such a diversion will not take place insofar as annual exports of pipes and tubes to the US do not exceed the 1979-1981 average share of annual US apparent consumption. In the light of its market forecasts the European Economic Community believes that exports of pipes and tubes to the US will not exceed this average. The EC expects that in these circumstances US steel producers will withdraw all pending countervailing duty petitions involving EC exports of pipe and tube to the US and will undertake not to file any petitions seeking import relief under US law including countervailing duty, antidumping duty, section 301 of the Trade Act of 1974 (other than section 301 petitions relating to third country sales by US exporters) or section 337 of the Tariff Act of 1930 on these products.

./.

Provisional address: Rue de la Loi 200, B-1049 Brussels — Telephone 735 00 40/735 80 40 — Telegraphic address: "COMEUR Brussels" —
Telex: "21 877 COMEU B"

139

- 2 -

B. The Community will establish measures with respect to exports of pipes
   and tubes from the Community to the US. Such measures will include communication to the US Department of Commerce of orders for exports to the US
   as shown in the order books of the European industry as of 1 October 1982.
   The measures will also provide for the Community to communicate to the
   Department of Commerce each month through 1985 the ex-mill shipments
   destined for export to the US.

C. Consultations may be requested at any time by the EC or US in the light
   of the market developments or in the event of any particular problem in
   trade between the EC and the US in pipes and tubes. In the context of
   consultations all statistical evidence that is available will be presented.

D. If estimates based on the above information and projections of US apparent
   consumption of pipes and tubes show that the 1979-1981 average described
   in paragraph A might be exceeded or that a distortion of the pattern of
   US-EC trade is occuring within the pipe and tube sector, consultations
   between the EC and the US will take place in order to find an appropriate
   solution. If after 60 days no solution has been found each party will
   take within its legislative and regulatory framework, measures which it
   considers necessary. In doing so both parties will act in a complementary
   fashion in order to prevent diversion.

E. If in any consultations held pursuant to paragraph D above it appears
   (based on substantial objective evidence such as allocation, extended
   delivery periods or other relevant factors) that the exceeding of the
   average described in paragraph A is due to supply or demand factors and
   that the US steel industry will be unable to meet demand in the US for
   a particular product then diversion shall not be considered to exist.

./.

- 3 -

F. If during the period in which the arrangement provided for in this
   exchange of letters is in effect, any petition seeking import relief
   under US law, including CVD, AD, Section 337 of the Tariff Act of 1930,
   Section 201 of the Trade Act of 1974, Section 301 of the Trade Act of
   1974, or Section 232 of the Trade Expansion Act of 1962, are filed or
   investigations initiated or litigation (including antitrust litigation)
   instituted with respect to pipe and tube products, and the petitioner
   or litigant is one of those referred to in paragraph A of the present
   exchange of letters or in Article 2 a) of the Arrangement concerning certain
   steel products, the EC shall be entitled to terminate the present exchange
   of letters after consultation with the US, at the earliest 15 days after such
   consultations.

   If such petitions are filed or litigation commenced by petitioners or
   litigants other than those referred to in the previous paragraph, or
   investigations initiated on pipe and tube products, the EC will be
   entitled to terminate this exchange of letters if during consultation
   with the US it is determined that the petition, litigation or investigation
   threatens to impair the attainment of the objective of this exchange of
   letters.

   These consultations will take into account the nature of the petitions,
   or litigation, the identity of the petitioner or litigant, the amount
   of trade involved, the scope of relief sought, and other relevant
   factors.

I should be grateful if you would confirm the agreement of your Government
with the foregoing.

Yours faithfully,

On behalf of the Council of the
European Communities

E. DAVIGNON
Vice-President of the
Commission of the European Communities

THE SECRETARY OF COMMERCE
Washington, D.C. 20230

2 1 OCT 1982

Vicomte Etienne Davignon
Vice-President of the European Communities
Rue de la Loi 200
1049 Brussels
Belgium

Dear Mr. Vice-President:

I am writing you this letter to record the agreement of the U.S.
government to your letter of October 21, 1982, which reads as
follows:

> "The Honorable Malcolm Baldrige
> Secretary of Commerce
> Washington, D.C. 20230   USA
>
> Dear Mr. Secretary:
>
> I am writing you this letter to record the results of our
> discussions on pipes and tubes:
>
> Arrangement on EC Export of Pipes and Tubes
> to the United States of America
>
> A.   It has been agreed during negotiations on trade in
>   steel mill products between the European Communities
>   (EC) and the United States (U.S.) that for the duration
>   of the Arrangement negotiated for those products
>   diversions of trade from steel products described in
>   Appendix B of the steel Arrangement towards pipes and
>   tubes should be avoided. The U.S. Government wishes
>   trade in the tube sector to be examined at this stage.
>   The Communities are of the opinion that such a diversion
>   will not take place in so far as annual exports of pipes
>   and tubes to the U.S. do not exceed the 1979-81 average
>   share of annual U.S. apparent consumption. In the light
>   of its market forecasts, the European Economic Community
>   believes that exports of pipes and tubes to the U.S.
>   will not exceed this average. The EC expects that, in
>   these circumstances, U.S. steel producers will withdraw
>   all pending countervailing duty petitions involving EC

- 2 -

exports of pipes and tubes to the U.S., and will
undertake not to file any petitions seeking import
relief under U.S. law, including countervailing duty,
antidumping duty, Section 301 of the Trade Act of 1974
(other than Section 301 petitions relating to third
country sales by U.S. exporters) or Section 337 of the
Tariff Act of 1930, on these products.

B. The Community will establish measures with respect to
exports of pipes and tubes from the Community to the
U.S.

Such measures will include communication to the U.S.
Department of Commerce of orders for exports to the
U.S. as shown in the order books of the European
industry as of 1 October 1982. The measures will also
provide for the Community to communicate to the
Department of Commerce each month through 1985 the
ex-mill shipments destined for export to the U.S.

C. Consultations may be requested at any time by the EC or
U.S. in the light of the market developments or in the
event of any particular problem in trade between the EC
and the U.S. in pipes and tubes. In the context of
consultations, all statistical evidence that is
available will be presented.

D. If estimates based on the above information and
projections of U.S. apparent consumption of pipes and
tubes show that the 1979-1981 average described in
paragraph A might be exceeded or that a distortion of
the pattern of U.S.-EC trade is occurring within the
pipe and tube sector, consultations between the EC and
the U.S. will take place in order to find an appropriate
solution. If after 60 days no solution has been found
each party will take, within its legislative and
regulatory framework, measures which it considers
necessary. In doing so both parties will act in a
complementary fashion in order to prevent diversion.

E. If in any consultations held pursuant to paragraph D
above it appears (based on substantial objective
evidence such as allocation, extended delivery periods
or other relevant factors) that the exceeding of the
average described in paragraph A is due to supply or
demand factors and that the U.S. steel industry will be
unable to meet demand in the U.S. for a particular
product then diversion shall not be considered to exist.

- 3 -

F. If during the period in which this Arrangement is in
    effect, any petitions seeking import relief under U.S.
    law, including countervailing duty, antidumping duty,
    Section 337 of the Tariff Act of 1930, Section 201 of
    the Trade Act of 1974, Section 301 of the Trade Act of
    1974, or Section 232 of the Trade Expansion Act of 1962,
    are filed or investigations initiated or litigation
    (including antitrust litigation) instituted with
    respect to pipe and tube products, and the petitioner
    or litigant is one of those referred to in paragraph A
    above or in Article 2a) of the Arrangement concerning
    certain steel products, the ECSC shall be entitled to
    terminate this Arrangement after consultation with the
    U.S., at the earliest 15 days after such consultations.

    If such petitions are filed or litigation commenced by
    petitioners or litigants other than those referred to
    in the previous paragraph, or investigations initiated,
    on pipe and tube products, the ECSC will be entitled to
    terminate this Arrangement if during consultations with
    the U.S. it is determined that the petition, litigation
    or investigation threatens to impair the attainment of
    the objectives of this Arrangement. These consultations
    will take into account the nature of the petitions or
    litigation, the identity of the petitioner or litigant,
    the amount of trade involved, the scope of the relief
    sought, and other relevant factors.

I confirm the agreement of the EC to the contents of this
letter. I would be grateful if you would confirm the
agreement of the U.S. government with the contents of this
letter.

Yours Faithfully,

Vicomte Etienne Davignon"

                              Sincerely,

                              Malcolm Baldridge

                              Secretary of Commerce

E   CEO Letter

Identical Letters Sent to Chief Executive Officers of Petitioning Steel Companies

October 21, 1982

Dear CEOs:

This letter is in response to questions you have raised with respect to the steel trade Arrangements reached between the United States and the European Communities (EC).

Entry into effect of the Arrangements are conditional upon, <u>inter alia</u>, U.S. Government receipt of an undertaking from all petitioners listed in appendix A of the Arrangement not to file certain petitions seeking import relief. Such an undertaking does not, of course, waive your statutory rights to file such petitions. It is a statement of your commitment not to file such petitions on Arrangement products or pipe and tube imports from the EC while the Arrangements are in effect.

If a petition is filed on an Arrangement product, Article 2 b) would apply. Because Arrangement products are defined in Article 1 as "exports to or destined for consumption in the U.S. of products described in Article 3 a) originating in the Community," Article 2 b) does not apply to petitions or litigation that will not affect U.S. imports of those products from the EC.

With respect to Article 7 a) of the Arrangement on certain steel products, the U.S. Government will not agree to increase the percentage limit beyond ten percent.

With respect to Article 8 of the Arrangement on certain steel products and Paragraph E of the Arrangement on pipe and tube, the determination would be made by the U.S. following a review of relevant information, including that obtained from U.S. producers and consumers.

With respect to Article 10, the U.S. Government will request consultations anytime that imports from Europe of certain steel products other than an Arrangement product exceed the EC 1981 market share of U.S. apparent consumption and any of your companies asks us to requests such consultations.

Forecasts of U.S. market demand for certain steel products and for pipe and tube will be secured from Data Resources, Incorporated, and will be used in connection with both the Arrangement on certain steel products and the Arrangement on pipe and tube.

- 2 -

The historical data on pipe and tube and the 1982 change in the U.S.
tariff schedule preclude precise identification of product
categories at this time. We will consult with the EC upon entry
into force of the Arrangements to identify relevant product
categories for the purposes of the pipe and tube Arrangement. The
consultation would follow a review by the U.S. of relevant
information, including that obtained from U.S. producers and
consumers. I have enclosed a list of the TSUSA items included in
the pipe and tube Arrangement.

To restate what I have told you personally several times, this
Administration is committed to effectively enforce these
Arrangements.

Sincerely,

/S /

Secretary of Commerce

Enclosure

F   Memo on "Price Increase"

UNITED STATES DEPARTMENT OF COMMERCE
International Trade Administration
Washington, D.C. 20230

October 14, 1982

MEMORANDUM FOR The File

From: F. Lynn Holec
             Director
             Agreements Compliance Division

Subject: Deletion of "Significant Price Increase" in Article 8 of
          EC-U.S. Steel Arrangement

On October 14, 1982, the Commission of the European Communities
acceded to the U.S. request to delete the reference to "significant
price increase" in Article 8 of the EC-U.S. Steel Arrangement. The
U.S. made this request because representatives of the U.S. steel
industry strongly objected to that reference in the August 5
Arrangement, fearing that a future U.S. administration would use the
threat of opening the gates to European imports as a means of
limiting steel price increases. The DOC and EC agree that changes in
price levels are an appropriate factor to consider in determining
whether a shortage of a particular product exists. Where
appropriate, we will consider changes in price levels as an "other
relevant factor" under Article 8 of the Arrangement.

G   Minute on Transition Period

AGREED MINUTE

This Minute records the understanding between the USG and EC Commission with respect to the fourth paragraph of Article 1 of the Arrangement.

The EC estimates that exports of Arrangement products during the period August, September and October 1982 will not exceed 968,000 metric tons. This level of exports would not be an aberration within the meaning of Article 1. This conclusion is based on the assumption that definitive export statistics will not be at variance with this estimate. If exports are not in line with this estimate, then the EC will adjust the export ceilings for the Initial Period to reflect excess exports.

_____   _____
EC   U.S.

H Request Letters: Section 626

DELEGATION OF THE COMMISSION OF THE EUROPEAN COMMUNITIES

The Head of the Delegation

October 21, 1982

The Honorable
Donald T. Regan
Secretary of the Treasury
Washington, D.C. 20220

Dear Mr Secretary

Pursuant to Arrangements between the European Coal and
Steel Community (ECSC) and the European Economic Community
(EEC) and the Government of the United States of America,
dated October 21, 1982, the ECSC and the EEC shall establish
measures with respect to exports, including export licenses,
for steel mill products specified under the Arrangements,
which are exported after October 31, 1982 and before
January 1, 1986, and are exported to, or destined for
consumption in, the United States. Recognizing Section
626 (a) of the Tariff Act of 1930, as amended by Section
153 of Public Law 97-276, I request, on behalf of the
ECSC and EEC, that you monitor or enforce these measures
by appropriate means under United States law, including,
where necessary, requiring the presentation of an export
certificate issued by appropriate authorities within the
European Communities as a condition for entry into the
United States of such steel mill products.

Yours faithfully,

Roy Denman

Roy Denman

2100 M Street NW Suite 707 Washington DC 20037 / telephone (202) 862-9500 / telex 89-539 EURCOM

149

THE WHITE HOUSE

WASHINGTON

October 21, 1982

Dear Mr. Secretary:

Pursuant to Arrangements between the United States
and the European Coal and Steel Community (ECSC)
and the European Economic Community (EEC), dated
October 21, 1982, the ECSC and the EEC have agreed
to establish export licenses or other export measures with respect to steel mill products specified
under these Arrangements, which are exported after
October 31, 1982 and before January 1, 1986, and are
exported to, or destined for consumption in, the
United States. Accordingly, I request under section 626 of the Tariff Act of 1930, as amended by
section 153 of Public Law 97-276, that you monitor
and enforce the measures taken within these Arrangements by requiring when necessary the presentation
of valid export certificates or other documents
issued by appropriate authorities within the
European Communities or their member states as a
condition for entry into the United States of
steel mill products from the European Communities.

Sincerely,

Ronald Reagan

The Honorable Donald Regan
Secretary of the Treasury
Washington, D.C.   20220

I   Antitrust Letters

DELEGATION OF THE COMMISSION OF THE EUROPEAN COMMUNITIES

The Head of the Delegation   October 21, 1982

The Honorable
William F. Baxter
Assistant Attorney General
   in charge of the Anti-
   trust Division
U.S. Department of Justice
Washington, D.C. 20530

Dear Mr. Assistant Attorney General,

    The Arrangement to be entered into between the European
Coal and Steel Community and the European Economic Community of
the one part and the United States of America of the other part,
will be implemented on our side by a Decision of the Commission
of the European Communities acting under Article 95 of the Treaty
establishing the European Coal and Steel Community (ECSC Treaty)
and a Regulation of the Council acting under Article 113 of the
Treaty establishing the European Economic Community (EEC Treaty).
This Decision and Regulation will enter into force after the conclusion of the Arrangement. These Acts will be published in the
Official Journal of the European Communities.

    The legal nature of this Decision and this Regulation is
as follows. According to Article 14 ECSC Treaty "decisions shall
be binding in their entirety" and according to Article 189 EEC
Treaty "a regulation shall be binding in its entirety and directly
applicable in all Member States". The Decision and the Regulation
will state that they are "binding in its entirety and directly
applicable in all Member States". As has been made clear by a
series of judgments of the Court of Justice of the European
Communities, this means that the Decision and the Regulation have
force of law in the Member States; moreover, such Decision and
Regulation take precedence over national law.

                                                    ./.

2100 M Street NW Suite 707 Washington DC 20037 / telephone: (202) 862-9500 / telex: 89-539 EURCOM

- 2 -

This Decision and this Regulation will set up a system of export restrictions and controls. They will require that exports may only take place with a licence. They will establish or provide for the establishment by the Commission of the quantitative exports limits for the different products and allocate out, or provide for the allocation by the Commission of parts of such product quotas between the Member States. The criteria to be applied by the Member States in their subsequent sharing out of their allocations between undertakings will be laid down in the Decision and the Regulation. Member States' authorities will deliver licences for each undertaking's shares. The Decision and the Regulation will prohibit exports to the U.S. without licences.

This Decision and this Regulation will provide that the Member States shall ensure that sanctions are applied in respect of all such exports effected without production of such a licence and in respect of any other breach of the provisions relating to such licences. This provision as part of a Decision and of a Regulation has also force of law. In addition and in any event the general principle of "Community loyalty" laid down in Article 86 ECSC Treaty and in Article 5 EEC Treaty requires Member States to take all appropriate measures to ensure the fulfillment of the obligations resulting from such a Decision and Regulation.

Herewith is a list established per Member State of national legislative or regulatory provisions relating to export restrictions and sanctions for breaches thereof. Typical sanctions include fines, forfeiture of goods and imprisonment, the specifics of which vary from Member State to Member State.

In the Commission's view limitations on the exports of steel to the United States, mandated by this system of export restrictions and controls by the ECSC and the EEC, including the sharing out by the Member States of their allocation to the undertakings and compliance by the undertakings would not violate U.S. anti-trust laws. The Commission would be grateful if the Department of Justice would confirm that it shares this view.

Yours faithfully,

Roy Denman

Enclosure

U.S. Department of Justice

Antitrust Division

Office of the Assistant Attorney General   *Washington, D.C. 20530*

OCT 2 1 1982

Sir Roy Denman
The Head of the Delegation
Delegation of the Commission
 of the European Community
2100 M Street
Suite 707
Washington, D.C.   20037

Dear Sir Roy:

This letter is in response to the request of the Commission
of the European Communities (the Commission), set forth in its
letter of October 21, 1982, for the views of the Department of
Justice on antitrust questions regarding measures now being
considered by the Communities, pursuant to discussions with the
United States Government, to restrain the export of certain
steel products to the United States.

The Commission has advised us that, pursuant to a Decision
of the Commission acting under Article 95 of the Treaty establishing the European Coal and Steel Community and a Regulation of the
Council under Article 113 of the Treaty establishing the European
Economic Community, it will set up a system of export restrictions
and controls for certain steel products destined for export to
the United States. The Communities will permit companies to export
to the United States only with a valid export license and will
establish or provide for the establishment by the Commission of
quantitative export limits for each covered steel product. The
export licenses will be allocated to individual companies by
each Member State according to criteria specified in the Decision
and Regulation, in accordance with the Member States' obligations
under Community law. The Member States will apply legal sanctions
for violations of the system of export restrictions; typical
sanctions include fines, forfeiture of goods, and imprisonment,
the specifics of which vary from Member State to Member State.
The Commission asserts that the Communities have authority to
establish such a system of mandatory export restrictions and to
cause the supporting sanctions to be enforced. The Commission
further asserts that the system will be binding under Community
law both on Member States and the enterprises concerned.

We understand the above advice regarding proposed actions by the Communities and Member States to constitute your representation that the proposed quantitative restrictions on the export of steel to the United States, and the allocation of those quantities among individual firms, will be imposed as mandatory controls by governmental entities acting within their sovereign powers. In such circumstances, the foreign sovereign compulsion doctrine would preclude liability under United States antitrust law for conduct compelled by those controls. We believe that U.S. courts interpreting the antitrust laws in such a situation would likely so hold.

Sincerely,

William F. Baxter
Assistant Attorney General
Antitrust Division

DELEGATION OF THE COMMISSION OF THE EUROPEAN COMMUNITIES

The Head of the Delegation   October 21, 1982

The Honorable
William F. Baxter
Assistant Attorney General
  in charge of the Anti-
  trust Division
U.S. Department of Justice
Washington, D.C. 20530

Dear Mr. Assistant Attorney General,

The Arrangement on pipes and tubes to be entered
into between the European Economic Community and the
United States Government provides that the Community will
establish measures with respect to exports of pipes and
tubes to the United States, including communication to
the United States Government of data on orders and shipments of such products destined for export to the United
States. The Arrangement also provides for consultations
between the Community and the United States concerning
market developments or trade problems in pipes and tubes.
If appropriate solutions cannot be agreed upon, the Arrangement provides that each government will take measures which
it considers necessary within its respective legislative
and regulatory framework to prevent diversion of trade into
pipes and tubes.

In order to obtain the information to be communicated
to the United States of America pursuant to paragraph B of
the Arrangement, a Regulation will be made by the Council
acting under Article 113 of the Treaty establishing the
European Economic Community (EEC Treaty). This Regulation
will enter into force after the conclusion of the Arrangement. This Act will be published in the Official Journal of
the European Communities.

                                                        ./.

- 2 -

The legal nature of this Regulation is as follows. According to Article 189 EEC Treaty "a regulation shall be binding in its entirety and directly applicable in all Member States". The Regulation will state that it is "binding in its entirety and directly applicable in all Member States". As has been made clear by a series of judgments of the Court of Justice of the European Communities, this means that the Regulation has force of law in the Member States; moreover, such Regulation takes precedence over national law.

This Regulation will require all exporters of pipes and tubes from the Community to the USA to supply the Commission with the relevant order book and ex-mill shipment information. The resulting information will be supplied to the United States Government, and only otherwise disclosed, in a sufficiently aggregated form so that no reported number reflects data pertaining to fewer than four European corporate entities.

As in the case of existing Regulations requiring the provision of statistical information, this Regulation will also provide that the Member States shall ensure that appropriate sanctions are applied in cases of non-respect of their obligations by the exporters concerned. This provision as part of a Regulation has also force of law. In addition and in any event the general principle of "Community loyalty" laid down in Article 5 EEC Treaty requires Member States to take all appropriate measures to ensure the fulfillment of the obligations resulting from such a Regulation.

In the Commission's view the provision by the exporters of this information to the Commission, mandated by this EEC Regulation and the subsequent transmission thereof in an aggregated form to the United States Government, would not violate U.S. anti-trust laws. The Commission would be grateful if the Department of Justice would confirm that it shares this view.

Yours faithfully,

Roy Denman

U.S. Department of Justice

Antitrust Division

Office of the Assistant Attorney General   *Washington, D.C. 20530*

OCT 2 1 1982

Sir Roy Denman
The Head of the Delegation
Delegation of the Commission
  of the European Community
2100 M Street
Suite 707
Washington, D.C. 20037

Dear Sir Roy:

     This letter is in response to the request of the Commission
of the European Communities for the views of the Department of
Justice on antitrust questions concerning its proposed arrangement with the United States Government on pipes and tubes.

     The arrangement provides that the European Economic Community
will establish measures with respect to exports of pipes and
tubes to the United States, including communication to the United
States Government of data on orders and shipments of such products
destined for export to the United States. The information will
be collected on a mandatory basis by the Community under a regulation with sanctions for noncompliance. No such data will be
supplied to the United States Government nor otherwise disclosed,
except in a sufficiently aggregated form so that no reported
number reflects data pertaining to fewer than four European
corporate entities. The arrangement also provides for consultations between the Community and the United States concerning
market developments or trade problems in pipes and tubes. If
appropriate solutions cannot be agreed upon, the arrangement
provides that each government will take measures which it considers
necessary within its respective legislative and regulatory framework to prevent diversion of trade into pipes and tubes.

     As a general matter, nothing in the United States antitrust
laws precludes the conclusion of government-to-government arrangements. Of course this arrangement does not and cannot provide an
antitrust exemption for conduct which would otherwise be unlawful.
We do not believe, however, that implementation of the pipe and
tube arrangement would necessarily include conduct violative of
the antitrust laws. In particular we believe that the provisions

for the collection of data and the communication of such data to
the United States Government, as described above, would not violate
United States antitrust laws. We cannot express any view regarding the antitrust implications of specific measures which might
be used pursuant to the arrangement, following upon consultations
between the Community and the United States Government, since
the arrangement does not specify the nature of such measures as
might be undertaken.

Sincerely,

William F. Baxter
Assistant Attorney General
Antitrust Division

Index

A

Advance use. *See* Flexibility provisions

Alloy steel products: US industry position on, 68, 75, 77, 81, 82; and priority list, 80; EC position on, 80, 83; resolution of issue, 83–4, 85, 86–7, 88

Alloy tool steel. *See* specialty steel

American Iron and Steel Institute (AISI), 9, 24, 74

American Metal Market, 30

Antidumping investigations: petitions filed by US industry, 11, 12, 14–15, 16, 25; initiated under TPM, 13–15, 21, 28; of 1982, 25, 27–28, 33–34, 35–37, 70–72. *See also* dumping

Antidumping law, US, 11, 14–15

Antitrust law, US: and VRA, 3, 6–8; and 1982 Arrangements, 52, 85, 96, 100

Armco Steel, 81

Arrangement products, 85–6

B

Baldrige, Malcolm: and TPM, 21, 22, 23, 24; initial position in 1982 dispute, 29–30, 33; in early negotiations, 37, 38–39, 52–54, 58, 59; and suspension agreements, 59; and August 5 Arrangement, 65, 74; and latestage negotiations, 81, 83, 107; letter to steel executives, 86–7, 92, 96

A (continued, right column)

Bars: cold-finished, 5, 49, 53, 59, 77, 88; hot-rolled alloy, 49, 52, 59, 77, 80, 83; hot-rolled carbon, 46, 59, 75, 80, 82–3

Baxter, William, 100–101

Belgium, steel industry of: and subsidization, 23, 24, 27, 56, 57, 72; and suspension agreements, 61–62; and dumping, 70

Bethlehem Steel, 12, 26, 68

Black plate, 68, 75, 77, 80, 83, 84

Brazil, 23, 27

British Steel Corporation: subsidization of, 56, 62–63, 72; and suspension agreements, 60, 61, 63; found dumping, 70; and hot-rolled carbon bar, 83; proposed slab exports to US, 86, 88. *See also* United Kingdom, steel industry of

Brock, William, 38. *See also* United States Trade Representative

C

Cabinet Counsel on Commerce and Trade, 37

Canada, 18, 23, 32, 45

Carter, President Jimmy: Administration of, 11, 12, 14, 15, 19; and specialty steel, 43

Carryover. *See* flexibility provisions

CEO letter. *See* Baldrige, Malcolm : letter to steel executives

Certain steel products, 85–6

Cockerill-Sambre, 24, 70. *See also* Belgium, steel industry of

Commerce, US Department of: and AD/CVD cases, 15, 23, 30, 33, 34, 36, 55–57; and TPM, 15, 19, 20, 21, 23; and 1982 Arrangements, 37, 44, 46, 60–63, 80; and specialty steel surge mechanism, 43. *See also* Baldrige, Malcolm, *and* Olmer, Lionel

Congress, US, 2–3, 7, 12, 89, 101–102

Consultations, as element of 1982 Arrangements, 51, 75, 77–78, 79–80, 81, 83–88

Consumers Union Law Suit, 6–8

Council on Wage and Price Stability, 12

Countervailing duty (CVD) law, US, 15, 16

Countervailing duty (CVD) investigations: initiated under TPM, 21, 23, 28; against pipe and tube, 41, 99, 103; of 1982: petitions filed, 25, 28; investigations proceed, 33–34, 35–36; deadlines' effect on negotiations, 48, 51–52, 53; determinations, 55–57, 72–73, 85

Critical circumstances, 71

Cyclops Steel, 26

D

Davignon, Etienne: in early negotiations, 30–33, 37, 38–40, 48–50, 52–54, 59; and suspension agreements, 60, 61; and August 5 Arrangement, 65; and late-stage negotiations, 79–80, 83, 107–8

Davignon Plan, 9

dePayre, Gerard, 37

DISC, 38, 57

Diversion, 83, 85–89, 103–4

Dumping: US steel industry allegations of, 9, 15; and exchange rates, 13–14, 15, 28; margins, interpretations of, 70–71. *See also* antidumping investigations, *and* antidumping law, US

Duration of 1982 Arrangements, 39, 45, 48, 51, 53. *See also* termination

E

Economic Development Administration, 13

Enforcement of arrangements: US government views, 39, 46–47, 53, 58, 89–90; EC views, 45, 47, 54; US steel industry views, 48, 52, 68–69; legal basis, 51, 58, 89–90; in August 5 Arrangement, 67; of pipe and tube arrangement, 104–105, 107–108, 109

Environment, protection of, 13, 14

Eurofer, 57

European Coal and Steel Community (ECSC): relation to EC, 3; steel products regulated, 31, 39. *See also* European Communities (EC)

European Communities (EC): and TPM, 15, 20, 22, 23; and early negotiations, 24, 30, 34, 46, 50, 54, 57; effect of US policy on, 38; and suspension agreements, 61–64; and August 5 Arrangement, 74, 78, 83; and late-stage negotiations, 80–81, 106, 108–109. *See also* Davignon, Etienne, *and* Paeman, Hugo

European steel industry: exports of, 1; and VRAs, 3, 5; efficiency of, 12; AD/CVD cases against, 12, 14, 15, 27–28; and TPM, 17–19, 22, 24; divisions within, 5, 27, 33, 38, 109; and 1982 negotiations, 50. *See also* Pipe and tube industry, European; European steel market; *and* individual countries, steel industry of

European steel market, 9, 11, 17–18

Exchange rates, 13–14, 15, 28

F

Fabricated steel products, 5

Fair value, 11, 13, 33

Flexibility provisions: in early negotiations, 46, 47, 48–49, 51, 52; in August 5 Arrangement, 67; resolved, 92

Forecasting, 45, 46

France, steel industry of: and TPM, 12, 18–19; and subsidization, 23, 24, 27, 55, 57, 72, 102–103; sheet exports to US, 34; and suspension agreements, 62, 63; and dumping, 70–71

France, government of, 38

Frenzel, Representative, 90

G

General Agreement on Tariffs and Trade (GATT), 28, 44, 57, 107

Geographical mix, 3, 5

Germany (Federal Republic of), steel industry of: and VRAs, 3; and TPM, 18–19; and dumping, 70–71; relations with rest of EC, 24, 27, 33, 38–9, 56–57, 109; subsidies to, 55, 72; concern with immunization, 46; and suspension agreements, 62; and restraint levels of various products, 66–67, 80–81, 82–83. *See also* pipe and tube industry; of Federal Republic of Germany

H

Heinz, Senator, 89

Hoogovens, 19, 23. *See also* Netherlands, steel industry of

Holec, Lynn, 91

Horlick, Gary, 37, 74

I

Immunization, of EC steelmakers from US AD/CVD law: in early negotiations, 46, 47, 51–2, 53, 54; in August 5 Arrangement, 67; in late-stage negotiations, 74–5, 77, 80–1, 94–99

Inflation, 11, 12

Injury: and US AD law, 11, 15; in TPM, 13, 15, 22–3; and EC exports, 30–31; determinations in 1982 investigations, 35, 82, 85

Inland Steel, 26

International Trade Commission. *See* injury

Italsider, 70. *See also* Italy, steel industry of

Italy, steel industry of: and sub-

sidization, 24, 27, 55–6, 57;
and suspension agreements, 62;
dumping, 70
Issues Group, 34

J
Japan, steel industry of: exports
of, 1, 8; and VRA, 3; and
dumping, 11, 12, 21; efficiency
of, 12, 13. *See also* United
States, steel imports of; from
Japan
Jones and Laughlin Steel, 26

K.
Kaiser Steel, 80
Korea, South, 21, 23
Korf, Willy, 96

L
Lambsdorf, FRG Minister, 33
Linkage, of steel trade to other
international issues, 12, 14–15,
37–9, 58, 61–2, 63
Luxembourg, steel industry of,
62, 70, 71, 72

M
MacGregor, Ian, 63
Mannesman, 41. *See also* pipe
and tube industry; of Federal
Republic of Germany
Modernization, of US industry,
16
Multifibre Agreement (MFA), 46

N
Nails, 21
National security, 3, 11, 42
National Steel, 26, 68
Netherlands, steel industry of:

and subsidization, 56, 72; sheet
exports to US, 34; and retaliation, 38; and suspension agreements, 62; cleared of dumping,
70–1; and restraint levels,
66–7, 80–81. *See also*
Hoogovens
Nixon, Richard, 4

O
Olmer, Lionel: and early negotiations, 37, 38, 44–5, 48; in
late-stage negotiations, 74,
79–81, 86, 101–102, 107
Organization for Economic Cooperation and Development
(OECD), 32

P
Paemen, Hugo, 37, 44–5, 48, 80,
102
Pipe and tube arrangement: background, 31, 41–2; early negotiations, 39, 45, 48–9, 51, 52–4;
and August 5 Arrangement,
58–9, 65–6; late-stage
negotiations, 75, 99–109
Pipe and tube industry: European,
41, 42, 99–100, 102; of
Federal Republic of Germany,
41, 102, 105, 109
Plate, 45, 80, 82
Preclearances, 18–9, 21, 23
Product coverage: early negotiations, 39, 45, 46, 49, 52, 53, 59;
in August 5 Arrangement, 65–6,
68–9; US industry proposal,
75–8; resolution, 79–88
Product mix: in VRA, 3; in pipe
and tube arrangement, 103,
104–5, 107–8

Q
Quotas, 2, 12, 24–5, 38, 101–2

R
Rail: AD/CVD petitions filed, 26; in arrangement negotiations, 59, 75–6, 80–1, 83
Reagan administration, 19, 25. *See also* Baldrige, Malcolm; Olmer, Lionel; Commerce, Department of; *and* United States Trade Representative
Reagan, President Ronald, 23
Regan, Donald, 89
Republic Steel, 26
Restraint levels: Initial positions, 39–40, 44, 47, 48–9, 50; Early negotiations, 52–3, 55, 57–8; 59–60; in suspension agreements, 61–2; in August 5 Arrangement, 66–68; late-stage negotiations, 74–8, 79–88
Roderick, David: and early negotiations, 29, 58; rejects August 5 Arrangement, 68; and late-stage negotiations, 74–8, 81, 82, 83; and pipe and tube, 107
Romania, 23, 27, 34

S
Sacilor, 18, 72. *See also* France, steel industry of
Semifinished products, 59, 75, 80, 85–6
Sheet: coated, 66–7; cold rolled, 45, 67; galvanized, 47; hot-rolled, 45, 67, 81, 82
Sheet piling, 24, 75
Shortage provision, 47, 52, 67, 90–2
Sidmar, 70. *See also* Belgium, steel industry of

Slabs. *See* semifinished products
Solomon, Anthony, 3, 12
Solomon plan, 12–3
South Africa, 23, 27, 35, 72
Spain, 23, 27, 35
Specialty steel: in 1972 VRA, 5; initial US government position, 39; background, 42–4; EC positions, 45, 49, 59; US industry positions, 49, 52, 78; and August 5 Arrangement, 65, 68; excluded from Arrangement, 84–5
Stainless steel. *See* specialty steel
State Aids Code, 24, 27, 44, 47
Steel Tripartite Advisory Committee, 13
Strike: and shortage provision, 91–2. *See also* United States, steel imports of; and labor
Structurals, 45, 67, 68, 80, 83
Subsidies: as justification for US import protection, 3; and TPM, 16, 18, 19, 20, 21; to EC steel industry, 16, 18, 24, 56, 72–3. *See also* individual countries; alleged in 1982 petitions, 28; Commerce Department approach, 38, 55. *See also* countervailing duty investigations; countervailing duty law; State Aids Code
Surge Mechanism, 15, 21, 43
Suspension Agreements, 60–64

T
Tariff Act of 1930: new section 626, 89–90, 104–6, 107. *See also* enforcement; section 704 (c), 60–1
Tax incentive, 14

Termination: of basic Arrangement, *See* immunization; of pipe and tube arrangement, 104, 106, 107–8

Terne plate, 75, 80

"Third countries" in US-EC steel dispute, 3, 32. *See also* traditional market share

Thyssen, 70. *See also* Germany (Federal Republic of), steel industry of

Timetable of 1982 investigations, 27, 74, 85

Tin-free steel, 68, 75, 76–7, 80, 81, 83, 84

Tin plate, 52, 53, 59, 67

Trade Act of 1974: section 301, 8, 44, 96, 107; section 201, 42

Trade Agreements Act of 1979, 14

Trade law, US, 32. *See also* antidumping law, US; countervailing duty law, US; Tariff Act of 1930; Trade Act of 1974; Trade Agreements Act of 1979

"Traditional market share": and TPM, 24; in arrangement negotiations, 31, 45, 47, 55, 57; in August 5 Arrangement, 67; in final Arrangement, 86

Transfer. *See* flexibility provisions

Transition period, 58, 67, 92–4

Treasury, US Department of, 12, 15, 23

Trigger prices: calculation of, 13, 14, 15; and exchange rates, 13–15, 16; relation to US prices, 17–18, 20; relation to fair value, 18; ignored by EC producers, 20

Trigger price mechanism (TPM): first version, 12–16; second version, 15–26, 38–9

U

United Kingdom, steel industry of, 27, 38, 57. *See also* British Steel

United States Steel Corporation, 14, 15, 26, 84, 86. *See also* Roderick, David

United States, steel exports of, 1, 6

United States, steel imports of: pre-1977, 1–2, 3–5, 5–6; 1977 increase, 9; and labor, 1, 4, 91–2; from Japan, 3–4, 5, 32; from EC, 1–4, 9, 17–8, 19, 20, 22–3, 30–2, 56; and world steel demand, 4, 5–6; included in 1980 AD petitions, 15; of pipe and tube, 19, 41–2; in 1981, 20, 22, 30–1; covered by 1982 investigations, 34

United States, steel industry of: financial performance, 9; and TPM I, 12, 14; and TPM II, 21–26, 33; in early negotiations, 29, 44–5, 48; and pipe and tube, 71–2; and August 5 Arrangement, 65, 67, 68–9; and late-stage negotiations, 74–8, 81—2, 84

United States Trade Representative (USTR), 15, 37, 44

Usinor, 18–9, 70, 72. *See also* France, steel industry of

V

Voluntary restraint agreements (VRA), 2–8, 45

W

Washington Post, 30–1

Wire rod, 26, 46, 81, 83, 84

Y

Yugoslavia, 21

About the Author

Michael Levine is a graduate of Carleton College (Minnesota) and the Woodrow Wilson School of Public Affairs at Princeton University. He joined the Commerce Department in 1980 and participated in the events of this book from the reinstitution of the Trigger Price Mechanism to the conclusion of the 1982 negotiations. In 1983 the Commerce Department awarded Mr. Levine its prestigious Silver Medal Award in recognition of his contributions to the 1982 negotiations and to the development of countervailing duty policy. Mr. Levine now resides in St. Paul, Minnesota, with his wife Holly and son Sam, where he is a commercial banker with the First National Bank of St. Paul.